PRAISE FOR *NEURO DESIGN*

'With solid science as the starting point, Darren Bridger provides an eminently practical guide to designing for your customer's brain. *Neuro Design* is packed with actionable strategies and techniques and is a must-read for every marketer and designer.'
Roger Dooley, author of *Brainfluence*

'An interesting book on a topic which should be of great importance to anyone in the business of retailing, advertising and marketing. Darren Bridger deals with complex topics in an engaging and practical manner, covering all aspects of the interplay between brain function and product design. Such an understanding is crucial for ensuring consumers stop and buy rather than walking on by.'
Dr David Lewis, Chairman of Mindlab International and author of *The Brain Sell*

'Darren Bridger has created a super, easy-to-read book demystifying the world of neuro design. What he does very well is address the balance between the role of human creativity and the role of neuroscience in modern design. If you think neuro design is about creating bland designs by deconstructing beauty, you need to read this book. It's not about that at all. Darren introduces all the major themes underlying the science in engagingly written manageable chunks and introduces you to the key methods and tools. The text is peppered with interesting factoids that really help lubricate the subject matter. Any book that tries to explain the allure of memes has to get five stars from me.'
Jamie Croggon, Design Director, SharkNinja

Neuro Design

Neuromarketing insights to boost engagement and profitability

Darren Bridger

First published in Great Britain and the United States in 2017 by Kogan Page Limited

2nd Floor, 45 Gee Street
London EC1V 3RS
United Kingdom
www.koganpage.com

c/o Martin P Hill Consulting
122 W 27th St, 10th Floor
New York NY 10001
USA

4737/23 Ansari Road
Daryaganj
New Delhi 110002
India

ISBN 978 0 7494 7888 9
E-ISBN 978 0 7494 7889 6

British Library Cataloguing-in-Publication Data

A CIP record for this book is available from the British Library.

Library of Congress Cataloging-in-Publication Data

CIP data is available.

Library of Congress Control Number: 2016048162

Typeset by Graphicraft Limited, Hong Kong
Print production managed by Jellyfish
Printed and bound by CPI Group (UK) Ltd, Croydon, CR0 4YY

CONTENTS

ABOUT THE AUTHOR

Darren Bridger works as a consultant to designers and marketers, advising on using and analysing data that taps into consumers' non-conscious thinking and motivations. He was one of the original pioneers of the Consumer Neuroscience industry, helping to pioneer two of the first companies in the field, then joining the world's largest agency, Neurofocus (now part of the Nielsen company), as its second employee outside the United States. He currently works as head of insights at NeuroStrata.

ACKNOWLEDGEMENTS

I would like to thank the following for their help during the preparation of this book: Catarina Abreau, Neil Adler, Chris Cartwright, Chris Christodoulou, James Digby-Jones, Keith Ewart, Adam Field, Ernest Garrett, Oliver Main, Shaun Myles, Thom Noble, Christopher Payne, Juergen Schmidhuber and Kattie Spence. I would also like to thank Charlotte Owen and Jenny Volich at Kogan Page.

For Lucy

What is neuro design?

01

Figure 1.1 Some of the fields that contribute to neuro design

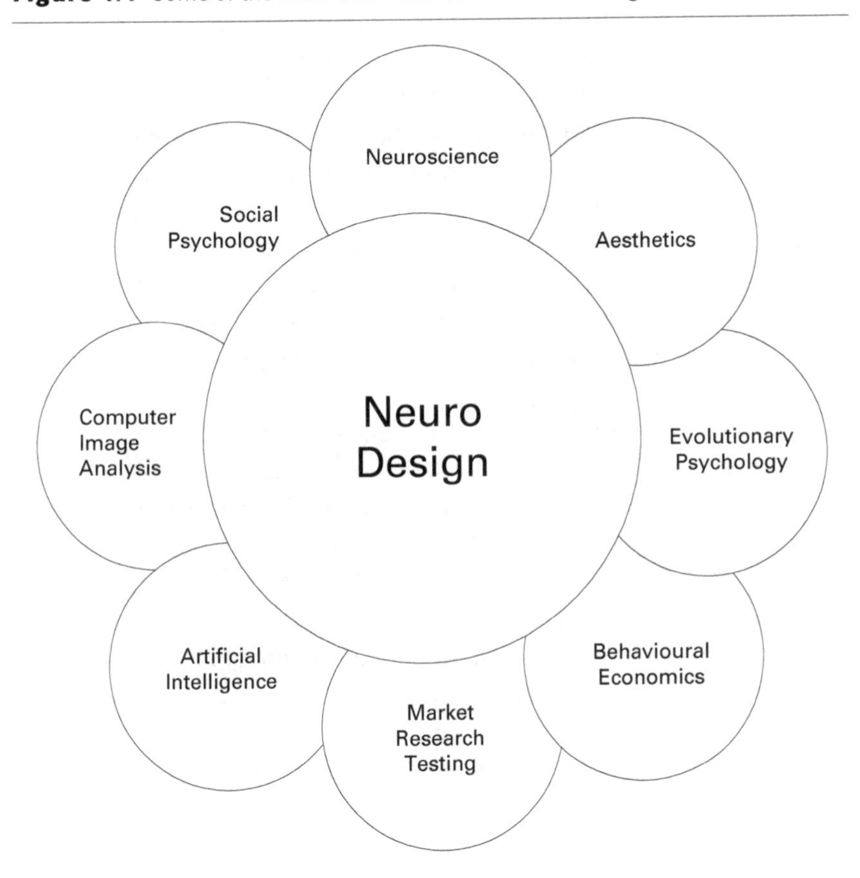

Sometime in the near future:

Dennis Drite sits down at his desk to begin a new day. Dennis – known as Den to his colleagues – works as a neuro designer, designing websites, ads and packaging. As he switches on his computer he logs on to his current design project – a retail website. As he finishes the design he runs it through his PNT software for checking. PNT, or predictive neuro test, checks his design, making predictions of how users will likely react to it. The tests for attention all come back positive, the images and text that are most important are all well designed to draw attention. Next the software tests for the fluency of the design: how easy users will find it to understand. There are a few minor issues here, and the software makes some suggestions for how to simplify the design to make it easier for people to understand.

The software then scans the face images he has put into his design. It measures the levels and types of emotions portrayed in the facial expressions. This feedback is useful because Dennis wants the faces to be emotionally engaging, and for users to feel some of those emotions. However, there's a problem with one of the faces: whilst the face is attractive, the software thinks people will find it bland. It morphs the shape and position of the face slightly, and it becomes more intriguing to look at. Dennis smiles approvingly and clicks to accept the change.

Finally the software runs a first impressions test. Dennis knows from research that, just as with people, our first view of a webpage design is critical for whether people will like it and stay, or click away within seconds. A surprisingly small number of design features are critical to making an effective first impression, and the software checks these and suggests a few small changes to the layout and colour scheme. The changes take seconds to implement, but will likely result in hundreds of thousands of extra users staying longer on the site and consequently buying more.

Dennis glances through the redesigned site. There are many more tests he could run, but these are the most important. The changes have resulted in a design that mainly came from his own creativity, but has been strengthened by the neuro software.

For years people had debated whether computers would take over from the work of human designers and artists. However, Dennis sees himself more like a Formula One racing driver: the marriage of human skill – his creative design instincts – with finely honed advanced machine – the neuro software. Indeed, on his wall he had hung a poster with a quote from Apple founder Steve Jobs: 'A computer is like a bicycle for the mind.'

Just as a bike boosts the power of our legs, the neuro software gears-up the strength of Dennis's own creativity. Dennis's design intuition combined with the feedback from the computer analysis creates a killer combination, a kind of *augmented intuition*.

However, intuition, theory and computer models only take Dennis so far; he wants to test the predictions with real human users. He clicks a button on his software and an invitation is sent out to a random selection of users of the site that says: 'Hey, we're testing our new website design and we'd love to see how you react to it. Would you mind us accessing your webcam so we can see where you are looking, and how you are reacting? We'll keep all our results secure and private and you'll be paid $1 for helping us.'

For those who click 'Yes', the video image of their face as recorded by their webcam is analysed to extract a number of reactions. Their eyes are tracked, to give information about where they were looking on the screen, moment by moment; their facial muscle activity is monitored to see even minute changes in facial expressions that give away their emotional reactions to the page; and even their heart rate can be monitored by the software, amplifying tiny fluctuations in their facial skin colour (imperceptible to the human eye).

Even in the earlier days of web design, designers and researchers had used A/B testing: showing different variations of a design to different users, and recording which group of users stay longer on the site, or are more likely to spend money. The effects of different designs were then inferred. However, whilst useful, A/B testing only took designers so far. Neuro design allows them to understand and predict which designs are likely to be more effective before they even exert the effort to create them.

What is neuro design?

This speculative foray into the possible future of design is my vision of the possibilities of neuro design. Whilst it may seem like science fiction, all the concepts in this scenario are based on real neuroscience research on how people view designs, and which elements make them effective. These concepts are new, and haven't yet all filtered down into all design degrees and courses, but they are likely to become an everyday part of the designer's toolkit in the near future. Already many design agencies are adopting them, or have at least realized their importance.

The software I described already exists today, although it is maybe not yet quite as sophisticated or as widespread in use as it will become. Nevertheless, the software itself is not essential to the process of neuro design, merely a way to automate it.

Neuro design is the use of insights from neuroscience and psychology in creating more effective designs. What can psychology and neuroscience tell us about what is really driving where we look in-store and online, what we choose to click on or pick up, what makes us share an image on a social network, and what image cues help drive a purchase decision? Neuro design also draws on other related fields to help build our understanding of why people react in the way they do to designs – fields such as computer image analysis (the ability of computers to analyse the composition of images, or even recognize what is shown in the image), behavioural economics (the study of how we make choices – often apparently irrationally – in spending our money) and evolutionary psychology (a branch of psychology that aims to explain behaviours in terms of how they would have evolved to help our ancestors survive).

Designers already use their own intuition in creating designs, and then in looking at the designs they have created to judge with their own eyes how 'right' they look, and making adjustments accordingly. They will also use a body of principles built up over the years amongst designers for how to create good design. Neuro design just adds to these principles. As neuroscience and psychology research have accumulated a lot of information over recent decades, they have many things to say about the common factors that influence whether people will like a design. This research has accelerated in recent years, including the founding of a specialized field – neuroaesthetics – that specifically studies the factors that influence whether our brains respond positively to images.

For the past decade I have been working with an eclectic mix of people, including neuroscientists, psychologists, marketers and designers, to develop insights into what factors make designs effective. Using these insights I have worked to help all kinds of companies – web design agencies, food and drink brands, automotive companies and film studios – to improve their design work. This work hasn't just been theoretical, I've worked with a new generation of neuroscience research tools that capture people's reactions to images and videos without having to use limited traditional techniques such as asking people to rate things on a 1 to 10 scale (we will take a closer look at some of these new research tools in Chapter 11).

You may have already seen some of the concepts of neuro design. Individual ideas are popping up on blogs, in journals and in books. Yet the subject might seem confusing at first glance: a series of disconnected recommendations that don't form a unified approach to design work. The aim of this book is to show how these ideas – and many you probably will not have encountered before – fit into a framework. A new way of looking at design.

However, before we look more closely at that framework, it is worth considering why neuroscience and psychology should be so important to the art and creativity of design.

A global psychology experiment

In a sense, the world wide web is like the largest ever psychology experiment. A psychological marketplace where every day millions of designs, photos and images are uploaded, and tested against millions of behavioural reactions: clicks. The typical psychology or neuroscience experiment in universities is run on around 20 participants, and may take months or longer before it is published. In contrast, the web is working to a radically faster rhythm, running real-time tests, globally, 24/7. Metrics such as social media 'trending' lists measure the global psychological pulse, telling us what people are thinking, feeling and desiring around the world.

The web has flattened out the relationship between those who create images and those who view them. In the past, the images created by designers and artists were consumed in silence, whether by the gallery-goer or the magazine reader. Now the viewer talks back. Yet, as we will see later in the book, it is largely not the conscious mind of the user that is talking, it's their non-conscious mind.

The web is also teaching us new things about the human mind. As an example, consider user-generated content. As recently as a couple of decades ago, few pundits predicted the volume of content that millions of people would enthusiastically post for free. Even experts like Bill Gates back in the mid-1990s imagined that the web would develop into something more akin to multichannel TV, with interactivity limited to things like viewers being able to click on a dress worn by a soap-opera actress so that they could order it themselves.[1] The model was top-down: big organizations pumping out content for the masses to consume. Most experts did not foresee the idea that a whole encyclopaedia could be crowdsourced for free with the voluntary efforts of amateur enthusiasts. Now we take Wikipedia for granted.

To a great extent we are now all graphical content creators. Even those without blogs or social media accounts may still be creating visual content in work presentations. Selecting shapes, images, clip-art and so on to illustrate their reports. Even aside from the web, digital tools have democratized design to an extent. At the time of writing, estimates suggest that more than half of all photos ever taken were taken within the last two years. Inexpensive smartphone apps allow people to apply filters and image manipulation effects that were recently the exclusive dominion of professional photographers and designers with access to expensive software.

Yet this efflorescence of creative activity has not diminished the role of the talented and trained designer. Their skills are probably more important than ever. Great design is becoming critical to business success.

One side benefit of the immediacy of the web is that it is possible to experiment and test different designs quickly and easily. A/B testing – uploading two different design variants, letting some users see one, some see the other – allows for quick testing of design effectiveness. However, whilst click measurements are a good behavioural test, they only answer half the question. They give the 'what' answer – what works – but they don't necessarily tell you '*why*'. Without understanding the underlying mental principles driving design effectiveness, A/B testing can be a trial-and-error process. The missing puzzle piece is to understand the hidden mental processes in the mind that are driving the mouse clicks. This is what neuro design aims to provide.

The importance of digital images

Perhaps the main thing the web has taught us is how much people like images. The web is very visually driven and becoming more so. Research has repeatedly shown the beneficial impact of digital images. Articles with good images are more likely to be viewed. Social media posts with images are more likely to be shared. Indeed, social media networks that are based around imagery – such as Instagram and Pinterest – have seen explosive growth. Equally, images and photos are a vital part of Facebook and Twitter.

We are visual creatures. We didn't evolve to read, but we did to look at imagery. It is our most salient sense, and the one that takes up the largest real estate in the brain. Hence we are adept image consumers. We decode them quickly and easily. They allow us to quickly absorb meaning. They allow us to grasp quickly the gist of a page or post, and guide our decisions to navigate deeper or away altogether.

Images lure us into reading content, make it easier to follow instructions, and are more likely to make content go viral. Some of the research findings on the importance of images online include:

- People are 80 per cent more likely to read content if it is accompanied by a colour image.[2]
- Infographics receive three times as many likes and shares on social media as other types of content.[3]

- Articles with at least one image for every 100 words are twice as likely to be shared on social media.[4]
- Content on the image-focused site Pinterest is highly likely to be shared by users: 8 in 10 pins on the site are 're-pins'.[5]
- Instructions that are accompanied by visuals are three times more likely to be correctly followed than those without.[6]

As we will see later (Chapter 8), people find it harder to read text on screens than on paper, even if the screen is high definition. In contrast, looking at images feels more effortless.

Design also has a real impact on how we value things. Sites with a design optimized to what the brain likes will be more trusted and liked. Products and packaging with better designs increase the amount of money people are willing to pay for them. In advanced economies where it has become harder for products to differentiate themselves clearly on just quality or functional features, design becomes a more important driver of value.

Web users are intuitive, impatient and image-focused

The proliferation of pictures has created a type of visual bombardment. Whilst we evolved to decode information from our eyes, the volume and variety of artificial images and designs we are now exposed to daily are unprecedented.

There are more images and choices available to us than ever before. If we are even slightly bored or dissatisfied with a page online we can click away. Hence we have all become skippers and skimmers. Research shows that people don't read things thoroughly online, they glance and dip.

With so much visual stimulation available, it is perhaps unsurprising that therapists and psychologists report a rise in the prevalence of attention deficit disorder-like symptoms. Those who have grown up with the web, digital natives, already consume content differently than their parents – comfortably shifting their attention between multiple screens at once. Yet divided attention is generally weaker than focused attention.

Research by Microsoft amongst 2,000 people in Canada showed that the ability to sustain focused attention in the face of distractions had dropped to just eight seconds in 2015 – down from 12 seconds in the year 2000, before the explosion of online images, videos and mobile screens.[7] This diminished attention span was reported by *Time* magazine to be below the putative focusing abilities of goldfish! Similarly, research at King's College London showed that being distracted by incoming e-mails caused a greater drop in IQ points than being stoned on marijuana.[8]

In China there is a military-style boot camp for young people (mainly boys) suffering from internet addiction. The guests undergo a rigorous programme designed by psychiatrist Tao Ran, who claims that internet addiction 'leads to problems in the brain similar to those derived from heroin consumption'.[9] In the West, expensive clinics that traditionally treat those suffering from addiction to alcohol, drugs or gambling have added internet addiction to their list of advertised problems they can treat.[10]

One study asked volunteers to sit in a bare room for 15 minutes and just be alone with their thoughts.[11] The only possible stimulation in the room was a button that would administer an electric shock. So intolerable was the lack of stimulation that 18 of the 42 participants chose to self-administer electric shocks rather than be quietly alone with their thoughts. Men were more likely than women to choose to shock themselves (12 out of 18 men, compared to 6 out of 24 women). It wasn't even curiosity that drove the participants to shock themselves: in preparation for the study they had already experienced what the shock felt like, and all participants found it unpleasant enough to say they would pay money not to experience it again.

The study authors think their results reflect the inherent difficulty of humans to control their own thoughts. Without training in techniques such as meditation, people prefer to focus outwards on external activities. Web browsing may just be filling this innate need, rather than generating it.

Another study shows the pleasurable stimulation that novel images give to the brain.[12] Participants were positioned inside a functional magnetic resonance imaging (fMRI) machine, scanning their brain activity whilst they were playing a card game on a screen. They were first shown a series of cards, each of which had a particular reward value attached to it. Then in the game they had to select cards one at a time. Interestingly, when presented with a new card that they hadn't seen before, they were more likely to choose it over a familiar one with a known reward. A primitive area of their brain, known to process neurotransmitters associated with good feelings (like dopamine) – the ventral striatum – became active. The novel option felt good, even if it was a more unknown and risky choice.

Back in our evolutionary history, whilst familiar things would feel less risky, we also needed to explore new things. For most of history our ancestors lived as nomadic hunter-gatherers, constantly needing to explore new territories in search of new sources of food. This pressure would likely have evolved in us a drive to see and explore novel things: one that is now showing up in a completely different context – in this example online browsing – yet taps into a primal, almost instinctive form of pleasure-seeking.

Dealing with all the choices and information available to us has become like trying to drink from a fire hose at full pressure. We have a keen thirst to imbibe, but we can only do so if we have ways of filtering the incessant jet. These filters are in our brains, and we will look more closely at them in the following chapters. Attention is a psychological commodity, and neuroscience can teach us how it works (see Chapter 6 for more on this).

The explosion of information and choices puts more pressure on the psychological elements of a business, such as the designs of its website, products and communications. People usually do not have enough time or energy to fully research and consider every detail of what they are reading, hence images can help influence their judgements.

'System 1' business

When our brains are offered choices that are too complex to fully research and rationally compute, we fall back on our gut feelings. These gut feelings are often driven by mental shortcuts that our brains have evolved to enable us to act fast in the face of uncertainty. Some of these shortcuts relate directly to how we decode images; some relate to how we make choices, given any particular presentation of options.

Systems 1 and *2* were names coined by psychologists Keith Stanovich and Richard West, but popularized by the Nobel prize-winning psychologist Daniel Kahneman in his book *Thinking, Fast and Slow*.[13] The two systems of thinking do not refer so much to parts of the brain as processes that the brain uses. They are a convenient description rather than hard categories, as in everyday life we are continuously using a combination of these systems.

System 1 is the name given to types of mental processes that are effortless and operate non-consciously. System 1 is bad at logical and statistical thinking. For example, it does not seek out all necessary information about something before it makes up its mind, as it should if it were being completely rational. It tends to use pattern recognition and imperfect but fast rules of thumb, rather than deliberate and rational calculated reactions.

System 1 thinking is important with images and designs. Often images and designs are inherently irrational. They are not intended to have a completely dry, logical response. Rather they might produce an intuitive gut feeling, or even an emotion. Often there is no logical reason to prefer one design over another, yet we do, and it can sway our choices of what to buy or what to give our attention to. Even designs that we think of as purely conveying functional information will have a feeling associated with them.

Psychologists have long recognized the importance of imagery to our non-conscious minds. Images have a primal access to our minds; before we learn to speak or read we can decode images.

System 2, in contrast, is slower and more effortful. If you were trying to calculate the value for money of two different food products by comparing their volumes and prices, you would be using System 2. As it is slower and takes mental energy and effort to employ, most people tend to shy away from using it where possible. System 2 is generally only triggered when we are unwilling or unable to use System 1.

As we encounter more images and designs in our daily lives than ever before, System 1 processes become more important. Yet as they are largely non-conscious, they are by definition outside our awareness and difficult or impossible for us to describe. However, we like to think that we are rational beings, in conscious control of what we do. If we are driven or influenced to do something by our non-conscious System 1 mind, but we are not aware of the process that has influenced us, we will tend to construct a reason.

Ironically, the more visually complex and information-rich our daily lives become, the less we rely on our rational, conscious and logical minds, and the more we fall back on our intuitive, non-conscious and emotional minds. 'With the sophisticated mental apparatus we have used to build world eminence as a species,' writes psychology professor Robert Cialdini, 'we have created an environment so complex, fast-paced, and information laden that we must increasingly deal with it in the fashion of the animals we long ago transcended.'[14]

Neuro design

The fact that our non-conscious mind is processing what we see, and shaping many of our reactions to it outside our conscious awareness, means that just asking people what they think of something is not good enough. Much of the time people are not consciously aware of the mechanisms in their brain that have caused them to prefer one design over another. Yet rather than say they don't know, people have a tendency to confabulate, that is to create a plausible-sounding explanation for their choice. We don't sense our own lack of insight, so we find it easy to 'fill in the gaps' with conscious rationalizations for our behaviour and choices. An example of this is an effect that psychologists have long known about, called choice blindness.

As the Digital Revolution has made images more important and prevalent in our lives, it has also advanced our ability to measure non-conscious reactions. Over the past few decades academic neuroscientists and psychologists have been developing a range of new techniques for measuring how our System 1 brains work. This has yielded new insights into how people respond to images and designs, but has also given us new tools for research that have become cheap and accessible enough to be used or commissioned by non-experts.

For example, eye-trackers can monitor where users are looking on a screen, measuring which aspects of a design get most attention and in what order. Eye-trackers have been used in commercial research for years, but the technology has dropped in price enough to be more affordable to a wider range of people and businesses. Also, eye tracking can now be conducted over the web via users' own webcams.

Webcams can also be used with a measure called facial action coding (FAC). The camera is used to capture the user's facial muscle activity and, from this, one of seven emotions can be inferred (there are seven universal facial emotions that seem to be present across all cultures). A new measure that can also be derived from webcams is Eulerian video magnification. This involves magnifying the minute changes in facial skin colour as the person's heart beats, deriving a measure of heart-rate changes.

Other online measures include implicit response tests. These are measures based on reaction speed, developed in universities to measure social biases that people may not be aware of or willing to consciously report. These tests work by giving participants a sorting task, quickly sorting words or images (such as logos) into one of two categories. However, before each image or word appears, an image (such as a design element or ad) will briefly pop up. This brief exposure will affect the speed of response on the sorting task, and from that the person's non-conscious reaction to the image can be inferred. The tests are versatile and have many benefits over more conscious-type questionnaires.

Other, more 'lab-based' measures exist that involve placing sensors on a person's head or body: for example, electroencephalography (EEG), which measures patterns of electrical activity in the brain by placing sensors around the head (usually fitted into a cap, a little bit like a swimming cap). This can give moment-by-moment measures of things like attention, and whether the person is emotionally attracted to an image. Another measure, called galvanic skin response (GSR), or electrodermal response (EDR), involves measuring changes in the electrical conductivity of the skin by placing sensors on the person's fingers.

The findings discussed in this book come from these types of measures, and other tests that are used by neuroscientists. These measures are also now in everyday use globally in evaluating designs of all kinds, from webpages to print ads, to package design. Typical questions that neuro design research addresses can include things like:

- How can this webpage design be changed to improve the first impression that users have of it?
- Which of several print ad designs is most likely to create the emotional reaction desired?
- How can a print ad design be optimized to make sure that it gets seen, and that the most important elements on it are seen?
- Which package design will most likely get noticed on-shelf in-store?

Using these measures allows you to test your designs against real human reactions. However, it is not necessary to use these to benefit from neuro design. The principles, ideas and best practices in this book can be used both in the creation of designs and in the development and critique of them.

Some of the principles of neuro design include:

- **Processing fluency**
  Our brains have a bias towards imagery that is easy for us to decode. Simple images or images that are easier to understand than we would expect have an advantage over more complex images, and these effects operate outside viewers' conscious awareness (see Chapter 3 for more on this).
- **First impressions**
  Our brains cannot help but make fast intuitive judgements when we see something for the first time. The overall feeling that results from this will subsequently bias how we react to the design. The surprising fact about this effect is that it can happen before we have even had time to consciously understand what we are seeing.
- **Visual saliency**
  As our brains are making sense of what we are seeing around us they construct what neuroscientists call a *saliency map*. This is a visual map of whatever our brain thinks is worth drawing our attention to. The interesting thing about images or elements of images with high visual saliency is that they can – like first impressions – bias our subsequent reactions. For example, research shows that package designs with high visual saliency can get chosen in-store, even when the customer actually usually prefers a competing product.
- **Non-conscious emotional drivers**
  Small details in a design can have a comparatively large impact on its ability to engage viewers emotionally. Creating an emotional effect is

important to making influential designs. Inbuilt biases in our brains can be followed to craft more emotionally meaningful designs.

- **Behavioural economics**
  In parallel to the research on how our brains react to images, a related discipline has grown up in the past two decades: behavioural economics. This relates to the way in which the quirks of our non-conscious can bias our choices, often in ways that seem irrational when they are exposed to us.

Many of the individual insights and recommendations from neuro design can be categorized under one of the five key principles above. By understanding the principles it helps to make sense of these individual insights, and prevents them becoming a hard-to-remember random list of recommendations.

Neuro design principles can be applied to almost anything that has an element of design, including webpages, logos, print ads, presentations and package designs. A good analogy is ergonomic design, where designers of products, buildings and furniture study the proportions and movements of the body in order to create usable designs. If ergonomic design is about understanding the shape, size and movement of the body in order to fit designs to it, neuro design is about understanding the quirks and processes of our non-conscious minds in order to create designs that appeal to it.

The interactionist approach

One question often asked about neuro design is: isn't design subjective and dependent on the particular tastes of the individual? Just as with food or clothes, we all have different preferences for what we find attractive to look at. The culture we have grown up in, and our experiences, can have a bearing on the development of these tastes. The opposing question is: is there such a thing as an inherently good or bad design? In other words: is good design in the design itself or is it merely in the eye of the beholder?

The approach taken in this book is referred to as the *interactionist* view. Whilst there are individual differences in taste, there are some common patterns of design that will be more effective across the population. Good design is about the use of these common principles, meeting the common features of most people's brains – an interaction of design and viewer.

This approach leaves enough space for the magic of the creativity of the designer. Design is not a science in which every element can be calculated and its effects entirely predicted.

Summary

- The web and digital screens mean that it is faster and easier than ever to test and measure the effectiveness of different designs.
- Image and design are particularly important on the web, as people use them to navigate and decide quickly when browsing.
- The explosion of images we all now see daily, and the choices available to us, put psychological pressure on us to filter them. This filtering is largely non-conscious and places a greater importance than ever on understanding how people non-consciously decode designs.
- System 1 thinking – which is non-conscious, fast and effortless – is crucial to understanding our reactions to designs.
- Neuro design is about insights from psychology and neuroscience into how our brains create different reactions to designs. It provides a series of principles that designers and those researching design can use to optimize their work.
- An interactionist approach to understanding design involves looking for common, repeatable design elements that have a particular effect on people. It assumes that good design is neither completely in the design itself, nor completely in individuals' interpretation, but an interaction of both.

Notes

1. Gates, B, Myhrvold, N and Rinearson, P (1995) *The Road Ahead*, Viking Penguin, New York.
2. http://www.office.xerox.com/latest/COLFS-02UA.PDF (last accessed 25 August 2016).
3. http://www.massplanner.com/10-types-of-visual-content-to-use-in-your-content-marketing/ (last accessed 25 August 2016).
4. http://buzzsumo.com/blog/how-to-massively-boost-your-blog-traffic-with-these-5-awesome-image-stats/ (last accessed 25 August 2016).
5. http://www.jeffbullas.com/2015/02/26/10-amazing-facts-about-pinterest-marketing-that-will-surprise-you/#UwHEGPES4LVJH9mf.99) (last accessed 25 August 2016).
6. Levie, WH and Lentz, R (1982) Effects of text illustrations: a review of research, *Educational Communication and Technology Journal*, **30** (4), pp 195–232.
7. http://time.com/3858309/attention-spans-goldfish/ (last accessed 25 August 2016).

8 http://nymag.com/news/features/24757/index6.html#print (last accessed 25 August 2016).

9 http://www.telegraph.co.uk/news/health/11345412/Inside-the-Chinese-boot-camp-treating-Internet-addiction.html (last accessed 25 August 2016).

10 For example, http://www.priorygroup.com/addictions/internet (last accessed 25 August 2016).

11 http://www.wired.co.uk/news/archive/2014-07/04/electric-shock-therapy-better-than-thinking (last accessed 25 August 2016).

12 http://www.world-science.net/othernews/080625_adventure.htm (last accessed 25 August 2016).

13 Kahneman, D (2011) *Thinking, Fast and Slow*, Macmillan, New York.

14 Cialdini, RB (1993) *Influence: Science and practice* HarperCollins, New York, p 275.

Neuroaesthetics 02

Figure 2.1 Palaeolithic cave paintings show systematic exaggerations of anatomical features of horses in order to make them more recognizable as having their summer or winter coats

In the mid-1940s a young American artist began painting in a barn, and ended up creating a revolution in the art world. Rather than positioning his canvas upright he preferred to lay it flat on the floor. Rather than brush paint, he preferred to flick and splatter it using sticks or pour it directly from the paint can. Rather than concentrate his movements on his painting hand, he moved his whole body, dance-like, with the paint tracing a continuous line around the canvas, tracing his rhythmic movements. The resulting paintings were apparently abstract, random lines of drips, yet to many were captivating and beautiful.

The painter was Jackson Pollock, and his drip paintings can now sell for more than US$100 million. His technique was later hailed as one of the greatest creative breakthroughs of the 20th century. In 1949 *Life* magazine asked, 'Is he the greatest living painter in the United States?'[1]

Yet why did such an apparently random, chaotic painting style create such an effect on viewers?

Just 52 years after Pollock made his first drip painting, a physicist called Richard Taylor discovered what he believes is the answer. In 1999 Taylor published a paper in *Nature* describing analyses he had performed on Pollock's paintings, showing that they contained hidden fractal patterns.[2] Fractals are patterns that are abundant in the natural world – they can be found everywhere from the human body to mountains and forests. Indeed most views of natural scenery will contain fractal patterns. Whilst they can look random they contain recurring patterns. For example, they have a property known as self-similarity. Different regions contain the same patterns, and if you zoom in you still see similar patterns to those you saw at full scale. It wasn't until the late 1970s and the Computer Revolution that mathematicians first discovered this hidden order in nature. Now when Hollywood special-effects artists need to design computer-generated realistic landscapes – forests, mountains, clouds etc – they can use fractal software. Computer fractal analysis has since been used to distinguish between genuine Pollock paintings and fakes, with 93 per cent accuracy.[3]

Creating a world with fractal design

In 1981 the computer graphics division of the Hollywood effects company Industrial Light and Magic (which later became Pixar) used fractals to create the first completely computer-generated cinematic scene. The sequence, for *Star Trek II: The Wrath of Khan*, depicted a barren alien planet being transformed into a living biosphere in the space of a minute, whilst the viewer orbits the planet from space then swoops down over its landscape. The shot required realistic landscapes, including natural terrain, shorelines and mountain ranges. Simulating such a planet was only possible thanks to software using fractal maths.

Taylor's work showed that Pollock's technique was not random. It was a precise depiction of fractals three decades before anyone knew what a fractal was. Pollock did not create the paintings consciously; he said the source of his inspiration was his non-conscious mind. Taylor believes that, as we are surrounded by fractal imagery in natural environments, we can learn, non-consciously, to recognize and appreciate it. He asked 120 people to look at a series of drip painting-like patterns, some of which were fractal, some of which were not: 113 people preferred the fractal patterns without knowing why.[4] Other research has shown that viewing fractal images of nature is effective at helping people to relax – perhaps an example of people feeling at home amongst natural environments, similar to those our ancestors evolved in. There is even a particular range of 'fractal-ness' in a natural image that seems to be optimal.[5] In other words, it is not just the fact that the image is of nature, but that it contains a certain level of fractal pattern.

So the evidence seems to be that an artist and viewers can create and appreciate patterns without necessarily having a conscious understanding of them. Artists are expressing their own non-conscious minds, and their work talks to the non-conscious minds of their viewers. Then it takes the non-conscious intelligence of a computer to discover what is going on!

Aesthetics and neuroscience

Taylor's analyses of Pollock's paintings, and the evidence that fractals seem to be perceived non-consciously, suggest a couple of things. First, that whilst computers are dryly logical and cannot perceive the emotional impact of art, they can nevertheless give us insights into how and why we might appreciate it. Some people might see the idea of computer analysis of art as reductionist, chopping an image into bits in order to analyse them, whereas humans view and appreciate the image as a whole. Yet the interesting thing about Taylor's research is that fractal analysis takes into account the whole of Pollock's painting. In a sense it is a less reductionist form of analysis. Second, that understanding the non-conscious mind can yield insights into why people like looking at certain images. It is sometimes not enough to simply ask people why they like looking at a particular image: they may not consciously know!

Jackson Pollock was just one example of the new abstract painters. The invention of photography in the 19th century had created new competition for painters. Whereas most painters in previous centuries had focused on trying to literally depict landscapes or people (albeit often with heightened beauty), cameras could now theoretically do this better. Painters strove to produce reactions and feelings via their work that could not be achieved through other media such as photography or writing. As painter Edward Hopper says: 'If you could say it in words, there would be no reason to paint.'[6] As we will see shortly, accounting for why people like looking at certain traditional paintings, like those of landscapes, might be a more simple exercise than explaining appreciation for more abstract imagery.

Aesthetics – the philosophy of art and beauty – has been around for millennia. Religious and philosophical beliefs drove thinking on aesthetics in the Classical and Renaissance eras. In the Classical era thinkers such as Plato saw the universe as having an inherent beautiful geometric order. They discovered that mathematics could be applied to understanding the visual world and to musical harmony. Then in the Renaissance era artists such as Leonardo da Vinci studied the mathematical proportions of the human body. They believed that if the beauty of music and the human form could be understood in terms of mathematical patterns, then perhaps these patterns could be used to create beautiful buildings and art. Beauty was inherent in nature's laws.

However, it was not until the 19th century that a more systematic testing of people's reactions to images was undertaken. For example, the 19th-century German experimental psychologist Gustav Fechner founded the field of experimental aesthetics: using scientific psychological research to try to understand or quantify what people find beautiful or appealing to look at. Fechner studied things like visual illusions and aimed to correlate the size and shape of paintings with their appeal to viewers. Whilst art and science may seem like opposite fields, the one place in which they overlap is the brain.

Why our visual system is now so well understood

Neuroscientists still do not fully understand many features of the brain. There are still many mysteries, such as what consciousness is and how it operates. However, our visual system is now fairly well understood. The visual or occipital cortex, positioned at the back of our heads, processes information from our eyes. There are two main reasons why neuroscientists have been able to make good progress in understanding how it works. First, unlike many other brain areas that do multiple things, the visual cortex is simpler in that it is dedicated to vision. This has made it easier to understand. Second, there is a very direct mapping of what we see and how our visual cortex processes it. As neuroscientist Thomas Ramsoy explains: 'If you see a particular pixel on a screen, that pixel is spatially represented in your brain. If you then present another pixel slightly to the right of this first pixel, the spatial representation of this pixel in your brain will be at a relative distance and angle that matches the real world. We... [say] that the visual system is *retinotopic*, meaning that there is a topographic mapping between the "real" world and the way that the brain processes this information.'[7]

Neuroaesthetics is born!

Neuroaesthetics is the application of insights from neuroscience to aesthetics – using our understanding of the brain and psychology to help explain why people like looking at certain images. It studies beauty in many areas, including music, poetry and mathematics, but in this chapter I concentrate on visual beauty. Art and design can create many other effects such as intrigue, admiration, persuasion etc. We will consider those in future chapters.

Neuroaesthetics is a very young field, only really recognized since the beginning of the 21st century. Yet it built upon earlier scientific work. For example, cognitive neuroscientists and psychologists have been studying visual perception for at least 100 years. Equally, evolutionary psychologists have provided theories that aim to explain the most popular forms of art as adaptations that were useful in helping our ancestors survive. Being able to recognize camouflaged predators quickly, find useful environments that could provide safety or food, and being able to spot the right coloured berries and fruits to eat would all have proved to have great survival value. Those amongst our ancestors who were better at these things would be more likely to survive and pass on their genes. Over millennia, evolution would thus have biased our brains to like looking at particular things.

However, we need to be careful in attributing aesthetic preferences to evolutionary pressures. For example, research shows that people rate those with longer legs (particularly women) to be more attractive. Some have speculated that this might be a sign that the person is healthier, having suffered no childhood malnutrition or illnesses that could have stunted the growth of their legs. However, research has shown that this preference – again, particularly with images of women – has changed through history.[8] In other words, it's easy to construct plausible-sounding evolutionary explanations for effects that may actually be cultural.

Humans have been drawing for at least 20,000 years longer than we have been writing. We know that humans have been creating art for at least 40,000 years (and, obviously, this is just the oldest evidence we currently have; humans may have been creating art for longer).[9] Art also appears to be universal, with people across all cultures making art in different forms. Whilst on the surface art does not seem to directly provide us with the things we need to survive (it can seem more like a luxury part of life), its ancient origins suggest it is tied to brain activity that was important to helping our ancestors survive.

For example, evolutionary psychologists have explained the popularity of landscape paintings as an evolved predilection for viewing scenery that would have provided an ideal living environment for our nomadic ancestors. Across nations, people prefer images of savannah landscapes that have trees with branches low enough to climb. Wild animals, and intriguing exploration-encouraging elements such as rivers curving out of view are also preferred. The savannahs of East Africa are where most of our ancestors evolved, so for evolutionary psychologists these visual preferences are a kind of remnant preference for these images.

Equally, our preference for looking at attractive faces is explained by evolutionary psychologists as a consequence of looking for mates with good genes. Certain physical features – such as a symmetrical face – are signs of a healthy and robust genetic profile.

Image preferences for things that would have rewarded our ancestors are, in a sense, comparable to our preferences for sugary and fatty foods: a craving for the pleasure of things that would have been helpful to our ancestors' survival. For this reason, these types of popular images have been called, somewhat disparagingly, visual cheesecake. Yet, as the best chefs know, even the humble cheesecake can be produced to a high level of skill and artistry. Some of the effects I describe in future chapters may seem, at first glance, to be simple and unsophisticated, yet their appeal can be as strong as that of a delicious cheesecake!

Neuroaesthetics also approaches its subject from different directions. For example, one direction is to study the types of art that people like. Another is to understand the more basic processes of visual perception. The first approach tends to place more emphasis on the whole of the image, the second on the parts. Another branch of neuroaesthetics aims to locate which areas of the brain are active when we are having different viewing experiences. Functional magnetic resonance imaging (fMRI) scanners have made it possible for us to see, in real time, which areas of people's brains become more active when they are viewing images in experiments. They work by tracking blood flow. As areas of the brain become active, they require more energy, hence blood flows to those areas to replenish their energy supplies.

Some of these findings may seem a little obvious. Such as when landscape images are seen, a brain region corresponding to our processing of places becomes active (the parahippocampal gyrus), or when faces are viewed, the fusiform face region becomes active.[10] Nevertheless, they begin to demonstrate how brain activity can be linked to our processing of images.

However, other findings have been more interesting. For example, research has shown that when people view images that they personally rate as beautiful, an area of the brain called the medial orbito-frontal cortex (mOFC) becomes more active. The more beautiful the person finds the image, the more this area becomes active. The association of this area with the experience of beauty is further supported by the finding that it also becomes more active when a person hears a piece of music that they find beautiful. In contrast, the amygdala and the motor cortex become more active when people experience images or music that they find ugly. (Interestingly, one speculation as to why the motor cortex – which controls movement – should become active when someone views an ugly image is that your brain is preparing to move you away from the ugliness!)

In one sense this finding is profound. After centuries of trying to measure beauty we now have an apparently objective, physical way to quantify it. Yet in another sense it could be seen as superficial. We can already simply ask a person if they find a picture beautiful. The brain scans just reveal a neural correlate of this. Equally, this finding alone does not really give us any insight as to why a person found the image beautiful. They do not tell us what mental processes or rules led to the mORC becoming activated. Yet it's a start. Further research in the years ahead could yield these kinds of insights. The well-known aphorism that beauty is in the eye of the beholder could perhaps now be updated to: beauty is in the medial orbito-frontal cortex of the beholder!

Whilst we don't yet have a full understanding of the brain processes behind why we find certain images pleasurable to look at, neuroscientists have begun to theorize some answers – using their knowledge of the brain. Two of the leading neuroscientists in the area so far are Vilayanur Ramachandran and Semir Zeki.

Ramachandran's nine principles

Vilayanur Ramachandran is an Indian neuroscientist who works at the University of California at San Diego. He has been one of the earliest and most influential contributors to the field of neuroaesthetics. One afternoon whilst sitting in a temple in India, Ramachandran came up with nine universal laws of art.[11] These are provisional ideas that have come from his knowledge of neuroscience and observations of art from around the world. He is not proposing that they are the only principles of how the brain perceives art, but they are a first draft of suggestions.

Several of these laws have a larger principle in common: the fact that when we recognize what something is, we can get a mini jolt of pleasure, an 'aha!' moment of insight. This ability of our visual brain goes unnoticed most of the time in our contemporary world. This is because we live amongst artificially pure colours and objects. For our ancestors, life on the savannah meant having to recognize camouflaged animals or objects obscured behind foliage. To be able to recognize a series of splodges of colour as the spots of a leopard was critically useful. Ramachandran's nine universal laws are set out below.

1. The peak shift principle and supernormal stimuli

One way in which art and design can heighten the 'aha!' moment of recognizing something is by exaggerating its most distinctive visual features. For example, caricature portraits usually exaggerate a person's most unique facial features: extending a long chin, growing a big nose, or inflating big ears. By exaggerating the unique elements of a person's face, the caricature cartoon can seem even more easily recognizable than a photograph of that person. Ramachandran evokes an ancient Sanskrit word – *rasa* – meaning the essence of something, to describe our brain's constant search for the most distinctive and identifiable visual elements of an object, person or animal. This type of design seems to mirror the way that our brains naturally process size differences. When people are shown two identical shapes, with one being slightly larger than the other, and then later are asked to draw them from memory, people tend to exaggerate the size difference. This suggests that we remember different features as simply 'larger/smaller' rather than remembering their size difference precisely.[12]

Peak shift is a term from studies on animal learning. For example, if animals are rewarded for learning to distinguish between two similar shapes – such as a rectangle versus a square – they begin to respond more strongly to more exaggerated versions of the rectangle than the one they were taught to recognize. When an animal is taught to respond to a stimulus, they usually respond most strongly (the 'peak' of their behaviour) to the exact form of stimulus that they have been taught with, but when they are taught to differentiate between two shapes, the peak of their behaviour shifts to a more exaggerated form of difference. What is happening is that the animal's brain is extracting the difference between the two shapes (a rectangle is an elongated square) and then responding more intensely to stronger versions of that difference.

An even stranger example of peak shift came from a study on herring gulls. The newborn gulls learn to peck at their mothers' beaks in order to ask to be fed. The mother's beak has a distinctive red dot on it, and it turns out that the baby gulls will also peck at a stick with a red spot on it. Their brains were simply responding to the red dot. However, researchers found that when they were shown a stick with three red stripes they responded more strongly – in a veritable frenzy of pecking! Somehow the three red stripes were like a super-stimulus to them: more intensely activating the link between a visual image and the pleasure of being fed.

The gulls reacting to the three red stripes might be a good analogy for the effect that art and design has on us. Whilst we evolved to visually recognize things that would help us survive or bring us pleasure, the brain rules for recognizing those things may often be based on simplified codes. By activating these codes – often simpler or unlike their real-world counterparts – artists and designers are effectively 'hacking' our visual system and stimulating it directly.

Even artworks that we think of as trying to literally represent something – such as a landscape painting, or a sculpture of a figure – often use exaggeration to create pleasing effects. This is a form of the peak shift principle: taking the elements that we find most interesting or useful in recognizing something, and heightening them.

Of course, a lot of art and design has already filled our world with amplified and heightened imagery that our ancestors would not have encountered in their environment. Peak shift images can be thought of as an example of 'supernormal' images.

For example, today we have the technology to display millions of colours on our screens, and a vast array of dyes and colouring agents for creating clothes, paints and products in many colours. However, our ancestors did not have so many pure versions of different colours. For example, the word for the colour orange did not appear in English until the 1540s, after fruits like oranges had started to be imported (even carrots were more brown, red or yellow before that time) – it was a colour that had not been seen frequently before then. The English term for people with orange hair is 'redhead', a term that dates back to at least the mid-1200s, before English people were frequently seeing the colour orange.[13] The visual world of our ancestors would have been comparatively drab and filled with less variety of colour and design than our contemporary world. Yet the love of supernormal imagery was present even thousands of years ago. Visual analysis of cave paintings has shown that their Palaeolithic creators systematically exaggerated the anatomical features of horses or bisons in order to make them more recognizable.[14] In other words they were making supernormal art.

The peak shift principle can be used in design in the following ways:

- If viewers are using a shape to find something, that shape could be exaggerated.
- The distinctive elements of photographs that make them appealing – such as the beauty of a landscape or the delicious appearance of food – can be exaggerated to evoke a stronger emotional response.
- Exaggerating characteristics of independent design elements can make them more distinct.

Exaggerating a face to make it more memorable

Researchers at the Massachusetts Institute of Technology have developed software that modifies photographs of faces to make them more memorable.[15] Whilst we are familiar with magazine designers airbrushing photographs of models to make them look more attractive or youthful, this algorithm instead tweaks elements of a person's face to make it look more distinctive. They researched facial elements that seemed to affect whether a person could remember a face or not, and incorporated the learnings into their algorithm. The resulting changes in the photographs are subtle but effective at making the face more memorable. In the near future, software like this could be used to help designers boost the memorability of all kinds of designs.

2. Isolation

Recognizing objects or people when they are partially obscured, or under imperfect viewing conditions (eg in the dark), requires mental effort. Thus when we see something under ideal viewing conditions it feels easier.

Isolation can be a bit like the peak shift principle: taking away all the visual features that are unnecessary for recognizing what the artist is trying to depict. We will explore this subject more in Chapter 3 on minimalist design. We will also look more closely at the idea of isolating a particular visual feature – such as movement, colour or form – later in this chapter.

How designers can use isolation:

- If something might be hard to recognize, avoid having other design elements overlap or obscure it.
- Selectively use white space around those design elements you wish to draw attention to.

3. Grouping

Whether we are putting together an outfit by selecting matching colours for the different items we are wearing or choosing a colour scheme for decorating our home, we find it natural to group things together visually. Our eyes merely take in the different hues and luminosities of light around us; it is our visual brain that must group these patterns together into objects and scenes. When apparently separate visual elements bind together in our minds, there is a pleasing 'aha!' moment. We can group things together for several reasons: because they move 'in sync' (eg the apparently individual dots we see are part of the coat of a moving animal); because they are of the same colour, similarly patterned; or because their lines and contours match up. There is even a neuro-correlate for this: when we recognize different visual elements as belonging together, the groups of neurons that represent each element begin to fire together in synchrony.

How designers can use grouping:

- Even if different image elements are not close to each other on a design, they can be associated together through use of things like colour and shape.
- Be aware that placing things together within a design can imply that they are related.

The Johansson effect

A dramatic example of how adept our brains are at grouping is the Johansson effect. If you take a person, put them in an entirely black bodysuit and mask, with white dots positioned over the body and limbs, and then film them moving so that on the film you can see only the movement of the dots, not the person, it is very easy to perceive the moving dots as a person. (Placing someone in an entirely black bodysuit with dots on may sound strange, but it is a technique used in the movie industry for motion capture: capturing the postures and movements of an actor in order to create an animated computer graphic version of them.)

4. Contrast

Things that have good levels of contrast are more easily recognizable. In opposition to the principle of grouping, contrasting colour combinations can be aesthetically pleasing because they jump out at us more strongly (a phenomenon called 'visual salience' that we will look at more in Chapter 6). Yet, similar to grouping, contrast helps our visual brain to spot the boundaries and contours of objects, aiding recognition. One study found that people's preference for natural landscape over urban images becomes reversed if the contrast of the landscapes is lowered.[16] Contrasts can also be conceptual: pairing images or patterns that we don't usually see together.

How designers can use contrast:

- If you want to draw attention to a design element, have it overlap a background or other design feature of a contrasting colour.
- Experiment with slightly raising the contrast of a design or photograph to make it more appealing overall.

5. The peekaboo principle

Our brains love solving simple visual puzzles. They are an artificial form of recognizing the leopard in the grass. By partially obscuring something it can make it more alluring. Recognizing something that is partially obscured is like solving a simple visual puzzle. Every day millions of people around the world willingly engage in puzzles for fun. There is fun in the seeking, but also in the resolution of the solution. Babies love it when adults play 'peekaboo' with them: alternately hiding their eyes behind their hands then popping out.

Visual perception, as we have seen, requires that we constantly form patterns out of the potentially confusing visual signals coming in from our eyes. Each time we manage to correctly recognize what something is, it is like a little brain-buzz. Of course, the puzzle shouldn't be too hard; there has to be an ideal balance between being just about hard enough to make the brain put in a little effort, but not so much that it is straining. Ramachandran believes that simple puzzle solving activates brain circuits involved in pleasure and reward.

How designers can use the peekaboo principle:

- Simple, easy-to-resolve visual puzzles can be a good way to capture attention and engage viewers.
- If an object or picture is already very familiar to viewers, can it be made more intriguing by partially obscuring it?

6. Orderliness

This principle relates to the regularity of an image. Do the lines and contours of a design line up? For example, if there are pictures hung on a wall we want them to be hung straight, not askew. Similarly, if there is a series of parallel lines in a design but one line is at an angle away from the others it can look wrong. This is similar to the principle of grouping: our visual brains have a strong drive to link things together. Whereas the real world is typically full of visual chaos, part of the pleasure of looking at art and design is that they can contain more regularity and orderliness. For example, repeating patterns in designs, like tiles that repeat a design, can be pleasing as the repeating of a pattern makes them easy to understand. Our brain only has to understand the small element that repeats, and then it understands the whole image.

How designers can use orderliness:

- If there are multiple lines of the same angle within a design, check before adding another line at a different angle: does it look wrong?
- Lining things up at the same angle can create a greater sense of balance and harmony in a design.

7. Visual metaphors

Visual metaphors are ways of mirroring an idea in visual form. Ramachandran gives the example of the way that cartoonists often use font types that mirror what the word means, such as words like 'scared' or 'shiver' written in a font that appears to be shaking. These types of design tricks help to re-enforce an emotion or the meaning of what is being communicated.

Visual metaphors can also be a type of visual rhyming or reflecting. For example, different elements within an image can mirror each other. Ramachandran believes that these types of metaphors work at a non-conscious level: we don't always consciously notice them. If we do discover them, it becomes like the peekaboo principle: we discover a hidden pattern, like solving a puzzle.

How designers can use visual metaphors:

- Can you borrow the technique of cartoonists in making your text mirror the meaning of the words?
- Look for ways to design images that act as metaphors for communicating concepts or emotions.

8. Abhorrence of coincidence

If something looks unlikely to have happened by chance, then it should make sense that it has been intentionally designed that way. Coincidences without reason look too obvious and somehow wrong. Our visual system uses something called Bayesian probability: essentially working out the likelihood of different interpretations of what it is seeing in order to create an interpretation.

Typically when we see something, our visual brain assumes that we are just seeing it from a random or generic vantage point, not a special vantage point that happens to make it look different.

Visual coincidences feel wrong because they are improbable. The only time they work is if there is a rationale for their use.

How designers can use the abhorrence of coincidence:

- If you are depicting an object or shape, check that the angle you are showing it from does not create any 'too convenient' visual effects.
- Beware of using orderliness or symmetry too obviously.

9. Symmetry

Symmetry, as we will see in Chapter 3, is pleasing because it makes a design easier to process. Another reason for why we like looking at symmetrical things is that in our ancestral environment it was usually a cue that we were looking at something biological. So noticing symmetry would have been a useful visual early warning for noticing predators, or for hunters spotting prey.

One watch-out with using symmetry is the previous principle of abhorrence of coincidences. If a design portrays something from a vantage point that makes everything line up in a symmetrical way, check that it does not look too obviously perfect. Of course, we are used to products – like cars or drink bottles – being symmetrical. We expect this, and it makes sense because these are objects that need to be convenient to use, and it wouldn't make sense for them to be too different on the right or left. For example, if you pick up a cola bottle, you don't want to need to think about whether you are picking it up from the left or right. However, imagery that does not have to be symmetrical can look 'too convenient' in some cases if it is.

How designers can use symmetry:

- Create symmetrical shapes, boxes and arrangements of images in your designs.
- If you are depicting something symmetrical, could you just show half of it? (Sometimes this can be a good, minimalistic way of conveying an image, as the other half is sometimes redundant.)

Summary

As we might expect with art, these principles are not absolute. If they are used it does not guarantee success. Also, sometimes breaking these principles can also lead to a pleasing image. For example, the peekaboo principle and the principle of isolation can be thought of as opposing ideas. The former suggests hiding an image in order to make the brain work a bit harder to discover the image, whereas the latter suggests making it easier to view an image. Similarly the principles of orderliness and symmetry argue for regular patterning, whereas the principle of abhorrence of coincidence argues for a more general randomness. Finding the right balance between these principles, or knowing when to evoke one rather than the other, is part of the designer's own skill, 'eye-balling' a design to see what looks right.

The principles are best thought of as a range of techniques – it's up to the designer to decide when it is appropriate to use each one.

The gestalt laws of perception

The gestalt psychologists were a 20th-century movement that examined how we view things as a whole. A lot of psychological research on perception is about breaking things into different elements and studying how we perceive them in isolation. In contrast, when we appreciate a good design or work of art we experience it in its totality. Gestalt psychology is concerned with how people perceive the whole as more than just the sum of its parts.

Although not strictly part of neuroaesthetics, gestalt psychology is a related area of thinking. For example, the principle of grouping that Ramachandran writes about is a big area of interest for the gestalt movement. They describe the many ways that our visual system groups different design elements together, using things like proximity and similarity.

The overarching principle of gestalt psychology is the law of *pragnanz*. This is the idea that we interpret images using the most simple and likely explanation. For example, in Figure 2.2 we assume that the shapes we are seeing are incomplete because they are overlapping each other (B), not because they happen to be incomplete in exactly the right way to allow their contours to match up (A).

Figure 2.2 Overlapping shapes

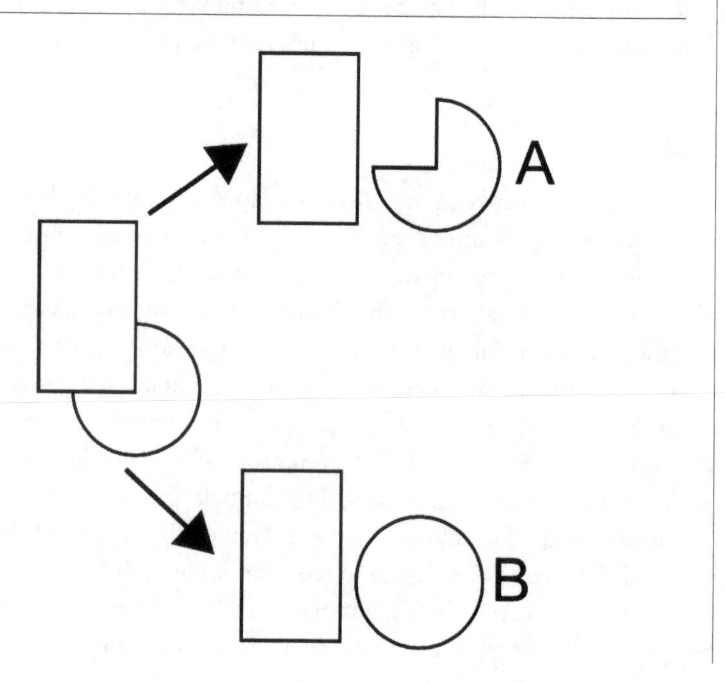

Again, this idea is mirrored by Ramachandran's principle of abhorrence of coincidence. When we group things together, we make the assumption that they are positioned in that way because they are intended to be seen as one whole, not that we are somehow just seeing an unlikely arrangement by chance.

Semir Zeki: artists are neuroscientists

Semir Zeki is a neuroscientist at University College London, with expertise in the visual system. He is one of the main pioneers of neuroaesthetics. In 1994 he co-authored one of the field's early papers, followed by one of its first books in 1999.[17]

One of his big insights is that modern artists have been doing something similar to visual neuroscience experiments without realizing it. There are a couple of ways they do this. First, artists non-consciously refine their work until it pleases their own brains. If their art also pleases the brains of others, then they have created something that the visual system likes.

The second way is a bit more complex. It relates to the fact that our visual cortex is organized into specialized areas (of which the main 'layers' are named with V numbers: V1, V2 and so on). In each area are cells that respond only to a particular facet of vision. For example, there are groups of cells that will respond to a particular colour of an object but are completely oblivious to an object's movement, orientation or shape. There are other cells that will respond to an object's orientation, but are completely indifferent to its colour, and so on. In order to discover that different areas of the visual brain are responding to very specific things, neuroscientists had to show people very specific types of images whilst scanning their brains. For example, images with no recognizable shapes, but with recognizable colours, or images in black and white, but with very specific shapes or line orientations. Zeki noticed that different styles of modern art seem to be doing something similar to these experiments. Whilst more traditional, representative art depicts recognizable people, objects or scenes, in many forms of abstract art the visual elements – form, colour, shapes and motion – are disconnected and treated separately from each other. Intriguingly this mirrors the way that our brains actually perceive the visual world.

For example, when we are judging the width of an object, we are unable to ignore its length. The two processes are connected in our brains. However, we do process the object's surface properties – such as texture and colour – independently from the form of the object.[18]

Equally, a style called kinetic art is focused on depicting movement. Zeki noticed that a number of prominent kinetic artists would minimize or eliminate colour from their work – almost as though they wanted to avoid activating the colour-sensitive cells – and concentrate their effects on just the movement-sensitive ones. He makes similar observations about cubist art, and notes that other artists, such as Cézanne, seem to be concentrating their effects on the orientation of lines.

Could artists in these fields somehow be aware of the separate processing of different elements like colour, movement and form? Intriguingly they could be.

As well as being processed in separate parts of the visual brain, these different visual elements also get processed at very slightly different times. For example, we first perceive colour, then form, then motion. The differences are tiny (measured in milliseconds) but in theory can be conscious. If artists become very sensitive to particular things (like colour or movement) maybe they are minimizing other visual features in order to isolate and tickle just the particular part of the visual brain that interests them. The more abstract schools of art may be just mirroring the fact that our visual brain separates out the processing of different elements of vision. This could even result in images that are intriguing to our visual system, even amongst those who are not consciously fans of abstract art. For example, those who do not consciously appreciate abstract art will often say things like 'My child could have drawn that!' Yet, interestingly, when their eye movements are tracked whilst viewing abstract paintings made by adult artists and those – with a superficial similarity – made by children, a different, non-conscious response is revealed. The paintings by abstract artists elicit greater visual exploration than those by children. There are more eye fixations, for longer and over more areas of the paintings.[19]

The implication of this for designers is that if they are trying to convey a particular visual element – colour, form, movement, line orientation – it might be best to dampen down variation in other elements. It may also be worth studying artists who have specialized in depicting that particular element. For example, Table 2.1 shows some examples of different artists and art movements that worked on isolating and amplifying colour, form or movement. Studying their techniques could be useful in finding ways to boost those features in designs.

Table 2.1 Art movements, the visual features they focused on and the visual brain areas they tickle

Art Movement	Visual Feature	Brain Areas
Fauvists (eg Henri Matisse, André Derain)	Colour	V4
Cubists (eg Pablo Picasso, Georges Braque)	Form	V1 and V2
Kinetic art (eg Alexander Calder)	Movement	V5

As well as his ideas on different artists mirroring the separation of how colour, form and movement are processed in the brain, Zeki has also proposed a couple of neuroaesthetic laws: constancy and abstraction.

1. Constancy

One of the most important things that our visual brain has to do is to recognize objects under different viewing conditions. For example, to recognize someone's face from an unusual angle or from afar, or to recognize the colour of an object even under extreme (low or high) light conditions. In the real world we have to do this quickly, easily and almost constantly.

Therefore our visual brain has to model an object or face in a generalized way – not just how the object or face looks under ideal viewing conditions or a particular angle and distance. This is similar to what Ramachandran refers to when he talks about *rasa* – the essence of something. Similarly, Zeki sees a parallel with this in art, as artists often try to capture the essence of something.

2. Abstraction

Do we view abstract designs differently from those that are more clearly representing something recognizable? As well as 'tickling' different areas and processes in our brain, abstract art may also be replicating something else that our brains do naturally. We often seek out pure, perfect examples of things. For example, we may long to see the perfect shape, or the perfect colour – in real life things are messy and have individual characteristics that do not always match an ideal. As part of the search for the essential visual features of things – which Zeki mentions in his law of constancy, and Ramachandran uses the word *rasa* to describe – our brains have to form ideal models. For example, we synthesize together all the faces we have seen to form an ideal model of what a face looks like. We cannot just remember every single face – we have to map them against a model (more on this in Chapter 3). So in our minds we have ideal models of things that we rarely actually see in the real world, but art can provide us with visual examples of these ideals.

Key concepts

Retinotopic

The description for the fact that the visual cortex directly maps on to the information from our visual field. This has made it easier for neuroscientists to understand the visual cortex.

Neuroaesthetics

Neuroaesthetics is the application of insights from neuroscience to aesthetics.

Gestalt psychology

A movement in psychology that seeks to look at how we perceive things as a whole. It is relevant to neuro design because it addresses how we link visual objects together.

Neuroaesthetics is starting to give us insights into how we perceive art, but its real impact will probably come in the years ahead. As we understand more about the brain, and are better able to correlate brain activity with the process of viewing and enjoying imagery, there will undoubtedly be more insights to come.

There are, however, rich arrays of insights from other areas of neuroscience and psychology into how we perceive images, which do not typically get organized under the label of neuroaesthetics. It is to those areas that we will turn in the next chapter.

Summary

- Neuroaesthetics is a relatively new field of study that uses neuroscience to study visual preferences or why we find certain things beautiful.
- Whilst appreciation of art is often a matter of personal taste, neuroaesthetics is usually concerned more with finding universal principles for appreciation of art.
- Some of the principles of neuroaesthetics relate to the idea that designs that help us to recognize what something is – by exaggerating distinctive features, increasing contrasts or isolating or grouping elements together – give us pleasure because they ease the effort of understanding.
- Many modern forms of art offer parallels with how our visual brain sees the world. Rather than being truly random or abstract, abstract art may actually appeal to people because it stimulates different modules of our visual brain. It mirrors how our brain visually decodes the world.

Notes

1. Jackson Pollock: is he the greatest living painter in the United States?' *Life*, 8 August 1949, pp 42–43.
2. Taylor, RP, Micolich, AP and Jonas, D (1999) Fractal analysis of Pollock's drip paintings, *Nature*, **399** (6735), p 422.
3. Taylor, RP and Jonas, D (2000) Using science to investigate Jackson Pollock's drip paintings, *Journal of Consciousness Studies*, 7 (8–9), pp 137–50.
4. Taylor, RP (1998) Splashdown, *New Scientist*, **159**, pp 30–1.
5. Taylor, RP (2006) Reduction of physiological stress using fractal art and architecture, *Leonardo*, **39** (3), pp 245–51 (the level of fractal-ness that appears optimal – most liked and best at reducing stress – is a D value 1.3 – 1.6; D is a measure of how fractal a pattern is).
6. Edward Hopper quoted in *New York Magazine*, 18 August 2013.
7. Ramsoy, TZ (2015) *Introduction to Neuromarketing and Consumer Neuroscience*, Neurons Inc. Holbæk, Denmark (Kindle book location: 1984).
8. Sorokowski, P (2010) Did Venus have long legs? Beauty standards from various historical periods reflected in works of art, *Perception*, **39** (10), pp 1427–30.
9. Aubert, M, Brumm, A, Ramli, M, Sutikna, T, Saptomo, EW, Hakim, B, Morwood, MJ, van den Bergh, GD, Kinsley, L and Dosseto, A (2014) Pleistocene cave art from Sulawesi, Indonesia. *Nature*, **514** (7521), pp 223–27.

10 Chatterjee, A and Vartanian, O (2014) Neuroaesthetics, *Trends in Cognitive Sciences*, **18** (7), pp 370–75.

11 Ramachandran, VS (2012) *The Tell-Tale Brain: Unlocking the mystery of human nature*, Random House, London. See also http://scienceblogs.com/mixingmemory/2006/07/17/the-cognitive-science-of-art-r/ (last accessed 25 August 2016).

12 Rosielle, LJ and Hite, LA (2009) The caricature effect in drawing: evidence for the use of categorical relations when drawing abstract pictures, *Perception*, **38** (3), pp 357–75.

13 http://allthingslinguistic.com/post/33117530568/why-dont-we-say-orangehead-instead-of (last accessed 25 August 2016).

14 Cheyne, JA, Meschino, L and Smilek, D (2009) Caricature and contrast in the Upper Palaeolithic: morphometric evidence from cave art, *Perception*, **38** (1), pp 100–08.

15 http://www.wired.co.uk/news/archive/2013-12/18/modifying-face-memorability (last accessed 25 August 2016).

16 Tinio, PP, Leder, H and Strasser, M (2011) Image quality and the aesthetic judgment of photographs: contrast, sharpness, and grain teased apart and put together, *Psychology of Aesthetics, Creativity, and the Arts*, **5** (2), p 165.

17 Zeki, S and Nash, J (1999) *Inner Vision: An exploration of art and the brain*, vol. 415, Oxford University Press, Oxford.

18 Cant, JS, Large, ME, McCall, L and Goodale, MA (2008) Independent processing of form, colour, and texture in object perception, *Perception*, **37** (1), pp 57–78.

19 Alvarez, SA, Winner, E, Hawley-Dolan, A and Snapper, L (2015) What gaze fixation and pupil dilation can tell us about perceived differences between abstract art by artists versus by children and animals, *Perception*, **44** (11), pp 1310–31.

Processing fluency

03

How to make designs feel more intuitive

Figure 3.1 Simple patterns in nature contain hidden geometric patterns that make them easy to look at yet interesting

The 1920s was the era of speed. Industrialization and new technologies like the radio, the phone and the car had both quickened the pace of life and impressed people with the greater wealth, efficiency and freedom available to us through scientific thinking. Industrialists such as Henry Ford were making new fortunes by adopting 'scientific management': timing how long it took people to perform work tasks and organizing factory production lines around their insights, and consequently seeing rises in productivity and hence profits. The success of scientific management implied that if we could study life rationally, we could all become more streamlined and efficient.

Design began to reflect this modern, rational ethos. A great 'spring cleaning' began, in which the more intricate or baroque designs from the previous century were swept away and replaced with cleaner, simpler-looking alternatives. Products, architecture and furniture became more streamlined and minimalist. The early years of minimalist consumer design in the 1920s and 1930s came from the European modernist movements, such as the Bauhaus design school. Their designs – embodied in everything from architecture to chairs – were almost shockingly modern for their time. Even today, some of the products and buildings from this era look surprisingly *futuristisch*. They convey a slick, rational and functional look, but most of all they are minimal.

However, whilst many people found these designs interesting to look at, their popularity waned: people generally found them too austere, cold and unfriendly to want to live with. Futuristic, functional simplicity was not enough for the public; they needed design that was more intuitively comfortable.

Raymond Loewy, the man called the 'father of industrial design', saved the idea of the modernist aesthetic.[1] In photos, Loewy – with his pencil moustache, slicked-back hair, suit and tie, and cigarette in hand – looks the very epitome of the mid-20th-century industrialist. Born in France, Loewy made his greatest impact after moving to the United States, designing everything from cars and trains to the interiors of President Kennedy's Air Force One plane, and Skylab, the US's first space station. However, it was probably in consumer design that Loewy became most influential. He took the streamlined look of modern planes, cars and trains, and domesticated it into consumer items such as lipsticks, fridges and radios. Loewy, in a sense, took the European aesthetic and made it more friendly and acceptable to consumers, popularizing the minimalist look.

Of course, updating the appearance of a product or brand is also a route to selling more, as it encourages consumers to update their products to stay current and fashionable. It also feeds our desire for novel things. As we saw in Chapter 1, we are often attracted to the novel, and get a small brain-buzz from seeing new designs. Yet if you push consumers too far beyond what they are ready for, this approach can backfire. It's about finding the right balance.

Loewy proposed a design principle he called MAYA, or *most advanced yet acceptable*. The idea is that the average consumer has expectations of what things should look like. These come from their prior experience with designs: we have expectations of what everyday items such as cars, phones or houses look like. Take phones, for example: 30 years ago a phone was an object tethered to a wall with a cord, with a hand-held receiver that you picked up. Today our idea of phone design is more likely to be a smartphone that we put in our pockets. Yet the smartphones that feel intuitive to us rely on a large body of knowledge that we have had to pick up and acclimatize ourselves to over the years – for example, the touch, swipe and scroll gestures to interact with the screen, the meanings of many different graphical icons, and typing with two thumbs on a flat surface without the tactile feedback of keys. All these things would have been unfamiliar to the public three decades ago, and would likely have rendered smartphones uncomfortably alien and complex devices to them.

In a sense, using advanced designs such as technological products and websites is like acquiring a language, or new set of behaviours. If consumers are pushed too hard or too fast, it feels too onerous and unnatural. The same is true of avant garde art and architecture. The paintings of Picasso and the architecture of Lloyd Wright were initially found ugly by many, even though they are now widely regarded as attractive.

Despite the original false dawn of Modernism, minimalistic design has been a strong trend in commercial design over the 20th century. Not just in the space-age sleekness of Loewy's designs, but the simple and clean designs of brands such as Apple, Ikea and Braun. Dieter Rams, the designer behind some of Braun's most iconic products, famously said: 'I believe designers should eliminate the unnecessary.'[2] In recent years, Apple's philosophy of Zen-like simplicity has been highly influential in mobile devices, helping to turn computing devices into more accessible, intimate parts of our lives.

However, whilst minimalism has been a strong trend in design, there is often the counter-pressure to include as much information as possible. For example, brand managers will often want as many benefits and claims as possible adorning their packaging and advertising; web designers will often need to include large amounts of information on their pages.

Equally, one of the conundrums faced by designers is whether to make their designs simple and easy to understand, in order to make them easy on the eye, or more complex and detailed, so that they are more interesting and involving. Another related conundrum is whether designs should be intuitively familiar and expected, or unfamiliar and *disruptive*.

This chapter is about ways to resolve these conundrums.

Processing fluency

There are obviously many different aesthetic styles that appeal to different individual tastes at different times and places. But there is something timeless and universal about the appeal of minimalistic design. This is because it appeals to the ways our brains decode and react to images.

Our brains are only a small part of our body mass, but eat up a lot of energy. So they have evolved ways to minimize the amount of energy they use, just like the power-saving modes on computers and household machines.

System 1, non-conscious thinking, is less effortful than System 2, conscious deliberate thought. As we saw in Chapter 1, our conscious System 2 minds have limited capacity. Whilst our non-conscious System 1 minds are constantly processing and sorting millions of bits of incoming sensory information, our conscious minds can only hold a few things in our awareness at once. Psychologists usually refer to this as our *cognitive load*. If we are trying to make a buying decision and there are many factors to take into account – such as weighing up prices, product features, how often we might use the product, whether the product is likely to be better quality than the other options, etc – it can overwhelm and overtax our conscious System 2 minds and thus the temptation to just let our System 1 mind make shortcut decisions can be overwhelming.

In general, when people are browsing websites, looking at ads or package designs, they don't want to have to consciously think too hard. They may be searching for a quick way to make a decision, for information or to be entertained. As we have seen, people are particularly impatient online and will lean towards designs that give them quick, easy and intuitive routes to what they are seeking. So it would make sense that simpler, easier-to-process designs might be favoured in this context. Psychologists call this 'processing fluency'. Information – such as an image or a task – that has good processing fluency is easy to look at or perform, and hence requires less energy.

We also have a preference for the familiar. In evolutionary terms, if we are familiar with something, it is generally not a threat. We evolved to live in small groups, and bonds of trust were built between those whose faces we were familiar with. We could generally feel comfortable with things and people we know – as they hadn't killed us!

Familiarity: the mere exposure effect

Professor Charles Goetzinger was a New Yorker teaching a class in communications at Oregon State University in 1967. Accounts of him describe his approach as somewhat eccentric, setting his students unconventional assignments, such as grading students on their ability to convince each other to sign a petition saying they should be awarded an 'A', and for one exam he merely instructed students: 'You have five minutes. Communicate.'[3] However, the strangest aspect of his classes was a mysterious student who turned up covered from head to toe in a black bag. Someone drove him to class every day and picked him up afterwards, maintaining his anonymity. Initially the other students reacted with hostility to the mysterious black bag character, but as the weeks passed they warmed to him, even becoming protective of him when the media appeared to report on his strange presence at the university.

The Polish-born psychologist Robert Zajonc heard about this incident and it intrigued him. He began to investigate the effect on people's emotions of repeatedly seeing something. He published his findings in a classic 1968 paper describing what he called the *mere exposure effect*.[4] Zajonc had performed a series of experiments in which different images, like simple shapes, paintings, faces and Chinese symbols were shown to participants in quick succession. Some of the images were repeated multiple times in the sequence, but because the images were flashing up so quickly it was impossible to consciously figure this out. Afterwards participants were asked which images they preferred and they consistently picked those that had been shown to them more than once. Just by exposing the images more often, people liked them more. The effect is important because it shows there is a non-conscious, non-rational mechanism in our brains that can lead us to like an image completely independently of any kind of logical evaluation of it.

One interesting fact about our non-conscious mind is that it gets confused between things that are easy to process and things that are familiar. The ease of looking at a simple image feels like the ease of looking at a familiar one. Familiar images – faces of people we know, for example – are easy for us to process, because we have already understood them. If we see something new, but it is easy for us to process, then it feels familiar and we like it. This feeling is generally not strong enough to make a lasting impression on us consciously, so we tend not to be aware of it.

A paper examining more than 200 experiments on the mere exposure effect found that it is reliable and robust, although tends to work best for brief exposures.[5] However, it is now believed that it is not exposure or familiarity per se that causes the preference, it is the fact that the more we see something, the easier it becomes to process (Figure 3.2). The mere exposure effect may just be a route to enhancing the processing fluency of an image. So we can group together simple images with familiar images, and complex images with unfamiliar images.

Figure 3.2 Familiar/easy to view versus novel/harder to view

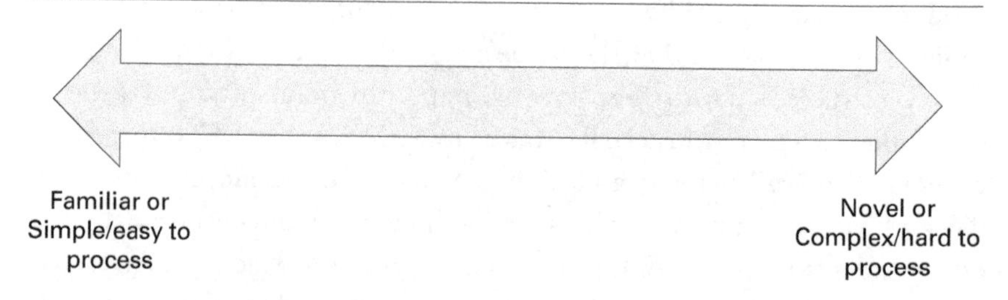

As well as studies showing that people express a choice preference for the familiar, there is also evidence that fluent images feel better to us.

Physiological evidence for processing fluency

A number of studies have used a technique called facial electromyography to test the emotional effect of processing fluency. Facial electromyography (fEMG) is a method for measuring emotions by placing sensors on a person's face to measure even tiny changes in the electrical activity of our muscles that make us smile or frown. If we have a negative emotional reaction to something, the corrugator muscle, which controls frowning, will become active. If we have a positive emotional reaction, the zygomatic major muscle, which creates smiles, will become active. If the theory is correct, images that are easier to process should activate the smile muscle; if they are hard then the frown muscle.

For example, in one study participants were fitted with the fEMG sensors and read a series of apparently random words.[6] However, some of the word lists had a common theme, whilst some did not. Table 3.1 gives an example:

Table 3.1 Word lists with a common theme are easier to process

List with a Common Theme (the Sea)	List with No Common Theme:
Salt	Dream
Deep	Ball
Foam	Book

The lists had been previously developed and researched, and the lists with a common theme were found to feel *coherent* to people even before they had consciously worked out what the common theme was.[7] When people were presented with the words with a common theme, their smiling muscle became activated and their frowning muscle relaxed. Other studies have shown similar results, with familiar versus unfamiliar female faces,[8] and dot patterns.[9] So there is evidence that processing fluency makes us feel good. Interestingly, they didn't find the opposite (frowning) for disfluent images, but we will come back to that shortly.

People internally monitor their ease of processing

Whilst we are largely unaware of our tendency to feel good about images that are easy to process, we can be aware of how easy or hard it feels to understand something. Our brains are monitoring how hard things are to process. We sometimes become aware of disfluent things, such as if we have to strain to read very small text, if a complicated image does not make sense to us, or if we glance twice at something because it seems out of context. Fluency is a 'pre-conscious' feeling: we are not always aware of it, but we can be if we direct our attention to it.[10]

People monitor their progress against an expected rate of progress and shy away from designs that demand too much thought from them. Unnecessarily eating up people's cognitive load should be a sin for a designer. Understanding a page or a task should be made as simple as possible, and the user unburdened, as far as possible, from the responsibility for figuring out. According to Apple's lead designer Jony Ive: 'True simplicity is derived from so much more than just the absence of clutter and ornamentation. It's about bringing order to complexity.'[11] In other words: the designer takes out as much of the hard working of thinking for the viewer as possible by doing it themselves.

The art of making complex information easy to process through clever design will become increasingly important, especially online. As Jony Ive suggests, this is not as simple as just stripping back a design so that it includes fewer elements.

However, simple or familiar images are not always universally preferred. Nor are we always unconscious of our reactions to images. Psychologists have begun to model how our System 1 and 2 minds can act in tandem to explain our reactions to images.

How Systems 1 and 2 decode an image

Laura Graf and Jan Landwehr have proposed a new model to help explain how we judge images both consciously and non-consciously.[12]

The 'pleasure/interest model of aesthetic liking' (PIA) starts with the idea that there are two things that influence whether we will like an image. First, whether it feels easy for us to process, and second, how we then (if we are interested enough) think about the image. The first is a System 1 process, the second a System 2 process. If people become interested in an image and they start to pay more attention to it, they can like it more. For example, when people are asked to look at and think about innovative car designs (prompted by a series of questions), they are more likely to like them.[13] Other research has shown that atypical car designs become more liked on repeated viewings.[14] Therefore a less familiar or more complex image can become more fluent as it becomes more familiar.

The model begins with a person looking at an image and finding it either fluent or disfluent. This results in an initial positive (fluent) or negative (disfluent) feeling about it. If the person is unmotivated to put more effort into thinking about what they are seeing then things end here, and they feel some level of displeasure with the image. This stage of the process is all System 1 and non-conscious.

However, if they begin to pay more attention to it – either because their curiosity is triggered or because they don't understand it and they need to, then several things can happen that will lead the person to either interest (if, after paying attention to it, it becomes clearer), or confusion/boredom.

The important thing is that the default way that people 'judge' images is with the System 1 feeling of fluency. It is only if they become motivated to learn more – for example if their System 1 feeling results in disfluency – that their System 2 attention can become triggered; it is a sign that they don't understand what they are seeing and they need to put more effort in to figure it out. For this reason, System 1 processing will often be more superficial, whereas System 2 processing will be deeper (Table 3.2).

Table 3.2 Summary of Systems 1 and 2

System 1	System 2
Automatic and effortless	More effort and attention needed
Driven by the image itself	More driven by the viewer's thoughts
Non-conscious	More conscious
The default mode for judging images	Triggered if an image doesn't make sense or we are motivated to understand it more
Mainly perceptual fluency	Mainly conceptual fluency
More superficial	More in-depth

Therefore, less simple or familiar images can still succeed, but only if people are motivated to understand them. However, as we will see below, there is also another way that images can become fluent.

Perceptual and conceptual fluency

There can be two types of fluency: perceptual and conceptual. The first is mainly due to its visual characteristics, the second to its meaning. For example, an image can be visually unfamiliar or complex, but convey a recognizable meaning, such as an unusual or detailed drawing of an object we are familiar with. The two can add together to form a general feeling of beauty/fluency. However, perceptual fluency tends to be more System 1 and non-conscious, whereas conceptual fluency is more System 2/conscious. An example of conceptual fluency would be the word lists (as shown in Table 3.1) and an example of perceptual would be a design that is easy to visually decode.

Propositional density

Simple designs can also convey rich meaning. This is called propositional density: conveying as much meaning as possible, with as little graphical detail as possible. Propositional density covers two elements: surface (the graphical elements) and deep (the meanings that the elements convey). For example, a surface element could be the use of the colour green, whereas the deep element would be the associations that green has with nature. The density can be expressed as a number: the number of deep elements divided by the number of surface elements. If the result is greater than one, the image conveys more meaning than just its basic graphic elements and is intriguing yet easy to view as a result.

Logos are a good example of images that often have high propositional density. For example, the Apple logo depicts an apple with a bite taken out of it, with the simplest of silhouette outlines and with just two graphical elements (the apple and the leaf at the top), yet can carry multiple deep elements of meaning. For example:

- Apples are natural and good for you.
- They are universal (for everyone).
- Sir Isaac Newton was hit on the head by an apple as the 'aha' moment that led him to form his theory of gravity (hence associations with intelligent insights).
- Depending on your culture, background and knowledge, the apple can also convey other things such as the fruit from the tree of knowledge; a slight association with outsiders and rebelliousness (eg upsetting the apple cart; or the myth of Eris, who throws an apple into a party to which she wasn't invited); or children giving an apple to their teacher.

Many meanings can be conveyed with the simplest of shapes.

Of course, brand logos can also naturally convey a number of meanings with just the simplest of design, simply because we learn to associate things with them through advertising etc. For example, the Nike swoosh is a very simple logo, yet it carries associations of athleticism, fitness, sports and so on. Other forms of meaning can simply come from our culture and upbringing (eg a simple image of an owl carries associations of wisdom, books, night time etc); or from natural shape associations (eg circles may convey unity, wholeness and inclusiveness; sharp or jagged design elements could convey the notion of something shocking or uncomfortable).

Images with higher levels of meaning can trump images that are merely simple.[15] For minimalist designs to be effective, they ideally should not be trivial, but information rich. As Leonardo da Vinci wrote, 'simplicity is the ultimate sophistication'.

Beyond simplicity versus complexity

Simple images are not always interesting, and complex patterns are not always easy to process. But a combination of the two can be both easy and interesting. Perhaps a better way to look at this conundrum is in terms of surface complexity versus information content.

One broader way to look at this is in terms of the surface complexity of the image – how much graphical information it contains – versus the information content of the image. By combining these two features of an image we can map our four main extreme types of image (see Figure 3.3):

1. **Low surface complexity, and low information content**
   This is something that is easy to look at, but conveys little meaning or hidden patterning. A basic shape like a circle would be an example of this. The problem with this is that it can be unsatisfying, and unless there is something particularly bold (such as use of colour) the use of the shape can just appear **bland** and meaningless.
2. **High surface complexity, and low information content**
   This would be something that has a lot of graphical detail, but no meaning or pattern behind it, such as the visual equivalent of white noise. The danger with this form of design is that it just appears **random**, and effortful to look at.
3. **High surface complexity, and high information content**
   This is a complex design that conveys a lot of information. The success of this type of design would depend on how motivated the viewer is to expend the effort to decode the complexity of the image. If they were not sufficiently motivated then the risk is that it will just appear **confusing,** and they will disengage from it.
4. **Low surface complexity, and high information content**
   This is the ideal type of image. It is easy to visually decode as it looks simple. Yet it also has a lot of meaning, or hidden visual information that a person can absorb at will and thus is **interesting.** Examples of this type of image would be a logo with high propositional density.

Figure 3.3 We can classify images, broadly, into four types based on their surface complexity and their information content (meaning)

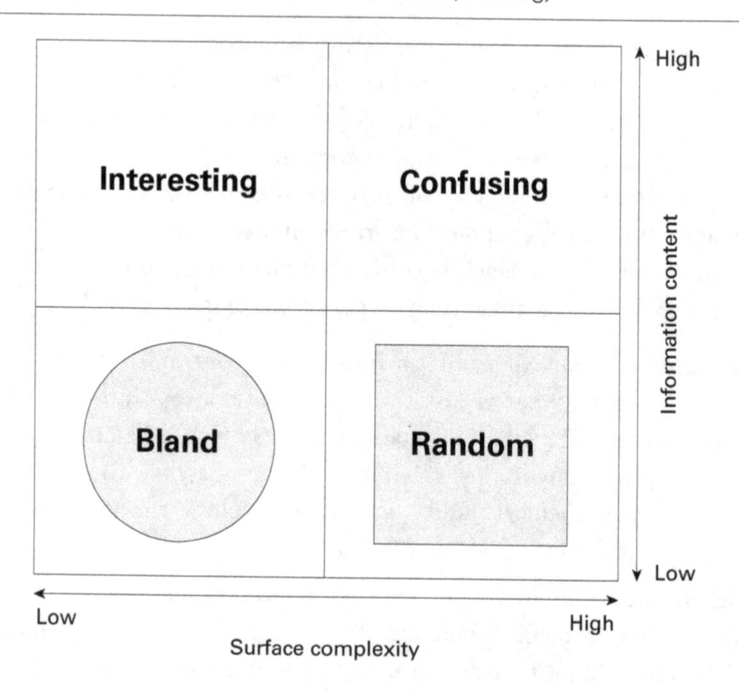

Therefore to be interesting and capture our attention, visuals should be as simple as possible on the surface but rich in information. As Edward R Tufte, data-visualization expert, writes: 'Graphical elegance is often found in simplicity of design and complexity of data.'[16]

Novelty and complexity can lead to liking

Making images simple is not always the only thing influencing whether they are liked. The experimental evidence shows that sometimes complex or novel images can also be liked. Novelty encompasses many pleasurable feelings: the antidote to the displeasure of boredom, the promise of new solutions to old problems, new jokes we haven't yet heard or new pleasures we haven't yet tired of. Yet novelty, being the opposite of familiarity, should be more *disfluent*. So what other factors can lead to novel images being liked? In other words, how can we understand the relationship between novelty and familiarity?

We have already seen that conceptual fluency can be more important than perceptual fluency. So a novel image could be liked if it carries a lot of meaning that is easy to decode. Equally, if a person is motivated, they may still like a more complex image if paying attention to it unlocks its meaning or makes it easier to understand.

Another way that novel imagery can be liked is based on our expectations of the image. If we expect something to be hard to understand, but it is presented to us in a way that makes it easy, we particularly like it. Things that are unexpectedly easy to process are like things that are unexpectedly familiar. For example, if you were on holiday and you bumped into a friend in a foreign country, you would probably find it far more pleasurable than if you bumped into them back home – it would put a bigger smile on your face. Making hard information easy to process mirrors this reaction.

The evidence now seems to show that it is not the ease of processing per se that leads to positive feelings, but rather how easy something is compared to how easy we expect it to be. In the fEMG studies described above, there wasn't evidence of frown activation from the more disfluent patterns, faces or word lists. However, as people generally expect things like faces, dot patterns and lists of short words to be easy to process, they were no more disfluent than expected.

This can be very dependent on the individual's own knowledge, and the context of what they are looking at. For example, in one experiment people were presented non-consciously with a series of Japanese Kanji characters (that they were previously unfamiliar with) that flashed up on a computer screen for just 13 milliseconds each.[17] Each one was repeatedly shown a total of 10 times in random order. The participants were then randomly assigned to one of three groups and were consciously shown a series of Kanji characters and asked to rate how much they liked each one, on a 1–9 scale.

In group one the participants were shown a mixed selection – half of the characters that they had seen before, half of them they hadn't. In group two they were just shown all the characters they had already seen. In group three they were only shown new characters that they hadn't seen before.

The mere exposure effect would predict that ratings of any characters they had seen before should be more favourable. However, what they found was that only when people were shown a mixed list, with unfamiliar characters mixed in with those they had seen before, were the previously exposed characters favoured. It suggests that the mere exposure effect might only occur when the familiar is in contrast with the unfamiliar.

This can help explain some of the lack of evidence from the physiological studies: people were probably already expecting these images to be easy to process. Therefore, despite being simple, they did not create a pleasing effect because there was no surprise in how easy they were to look at.

A good example of this is infographics. Infographics have become popular online in the last few years. They are one of the types of graphics that most often go viral online. Infographics, when done well, are information rich, often helping to make complex information seem easy to understand. Thus having unexpected simplicity.

So, people either prefer simpler than expected designs, or designs that convey a large amount of information in a surprisingly minimalistic way. However, not all designs have the potential to convey lots of meaning per se. Is there a way to still make them interesting to look at? One solution may come from artificial intelligence research.

How would you make a robot curious to view images?

Computer scientist Jürgen Schmidhuber works in artificial intelligence. There is a lot of ongoing work in this field on visual recognition: trying to make software that can take in inputs from its artificial eyes (cameras) and understand what is in front of it, just like humans do. However, ultimately, what is the good of an intelligent robot that can see and understand the world but has no motivation to explore and learn? Without motivation a robot is just a kind of slave: doing whatever it is told to do. Or it is dependent on external rewards: someone rewarding it every time it acts curious or explores its environment.

Whilst thinking about this problem Jürgen came up with an elegant theory that not only provides a model for how to make a robot curious, but may well explain exactly how we are motivated to explore and understand the world around us, and why we find certain images interesting and rewarding to look at.[18]

From a very young age, babies are like little scientists: curious about the world and seemingly inherently motivated to learn how things work. Even when we are older, we are still curious. Web browsing is a good example of this. A lot of web browsing is motivated by curiosity. The human urge to forage for information seems almost inherent in us.

The starting point for Jürgen was a fact that we discussed in Chapter 2: the inherent laziness of our brains. He calculated that, in theory, it is possible for our brains to store everything we see during our lives to the same level of quality as a DVD movie. However, storing everything and maintaining those memories take energy. Retrieving all those visual images from our memory also takes energy. What our brains needed was a shortcut. This is where the importance of learning general rules comes in. For example, when you look at a face, the actual visual image that your eyes receive will be different depending on the angle you see it from, the lighting conditions, etc. Yet we still learn to recognize it as the same face. What the brain is doing is storing a generalized pattern for that face so that it can recognize it under different viewing conditions.

Matching what we see to general visual memory patterns helps our lazy brains to save energy. Here is a simplified example. Take the simple grid image shown in Figure 3.4:

Figure 3.4 First grid pattern

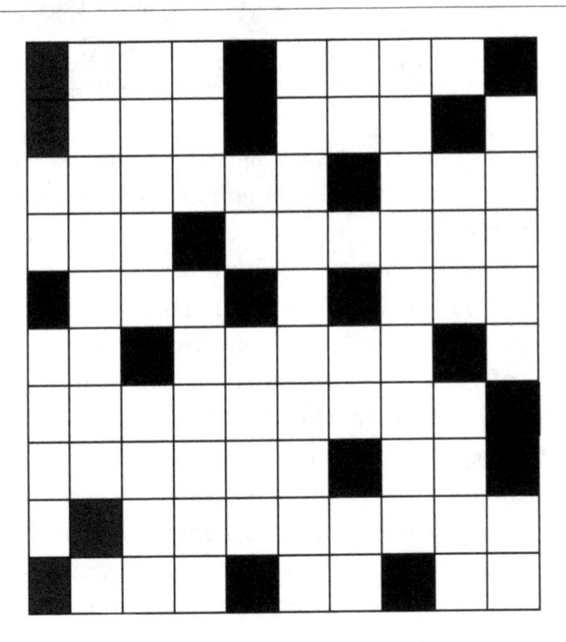

As there is no particular pattern to that image it is hard to memorize. You would have to remember every single cell that is filled in, one by one.

Now take a look at Figure 3.5:

Figure 3.5 Second grid pattern

As there is an obvious pattern to this grid, it becomes much easier to memorize. All you have to do is remember the pattern of black then white, repeated. You don't have to separately memorize each individual square.

In computer terms the image is 'compressible': we have compressed the image into a smaller thing to remember. In computer science and information theory the compressibility of a visual pattern is sometimes measured as its *Kolmogorov complexity*. Named after Russian mathematician Andrey Kolmogorov, this measure is based on the shortest computer program that could be written to reproduce the image. The shorter the program, the lower the Kolmogorov complexity and the more compressible the image. In theory an image could have a lot of visual detail, but if it is ordered in a repeating pattern of some form, it will have low Kolmogorov complexity. As long as the pattern is recognizable to our brains, it will make it easier for us to process too, once we have learned the pattern.

New compressible patterns drive our curiosity

Patterns are easier for our brains to memorize and process. They make images compressible, so our brain loves them and has an inbuilt bias towards finding them. This could be at the very core of our curiosity and motivation to seek out new information: we are hunting for compressible patterns.

This model explains why people have a preference towards finding those with similar faces to their own. Our brains store a pattern of the template of an 'average' face. This helps us to process and remember new faces, as we just then have to remember the difference between the new face and our brain's model of an average face.

This is an example of something called the 'beauty in averageness' effect. People tend to prefer average or 'prototypical' images. They are easier for people to understand and process as we already understand their 'template'. For example, in one study the researchers manipulated images of cars, birds and fish. They found that the more average versions of each image were preferred.[19]

How do our brains create a template for the average face? Simply from averaging out all the faces we see. And there are usually few faces we see more often than our own: we see it in the mirror daily. So our own face heavily biases our inner face template.

The process of our brains seeking to compress our models of the world down to simpler rules, and finding this process aesthetically pleasing, has a parallel in the work of scientists and mathematicians. When they find an equation that describes a lot of the natural world with just a minimal amount of factors, mathematicians and scientists describe it as having an elegance or beauty.

New compressible patterns are like a little snack for our brains!

The model also explains why we should find images rewarding to look at, even if they don't remind us directly of anything pleasurable. They are interesting for their own sake. We might not be consciously aware of why we like to look at them, but it is because it is feeding our brain with something it loves: new patterns that enable it to compress information.

There are many ways that images can provide us with new compressing patterns. For example, an image can contain geometric patterns, symmetry and regular proportions.

We do not necessarily need to be consciously aware of the existence of a pattern; it can be hidden. Our brains just have to be able to sense the promise of a pattern in order that they become interested in studying the image to extract the pattern.

Low-complexity design

As practical examples of his theory, Schmidhuber has created what he calls 'the computer-age equivalent of minimal art'.[20, 21] Low-complexity designs (and low-complexity art) are designs that at first glance may seem complex, but because they adhere to an underlying, regular pattern the image is 'compressible'. This makes them intriguing as we sense the patterned information they contain. As the pattern is regular, it is compressible/learnable, which makes the structure of the design easier to 'compute' (see Figure 3.6).

Figure 3.6 Example of low-complexity design: a face that 'fits' into a regular geometric pattern

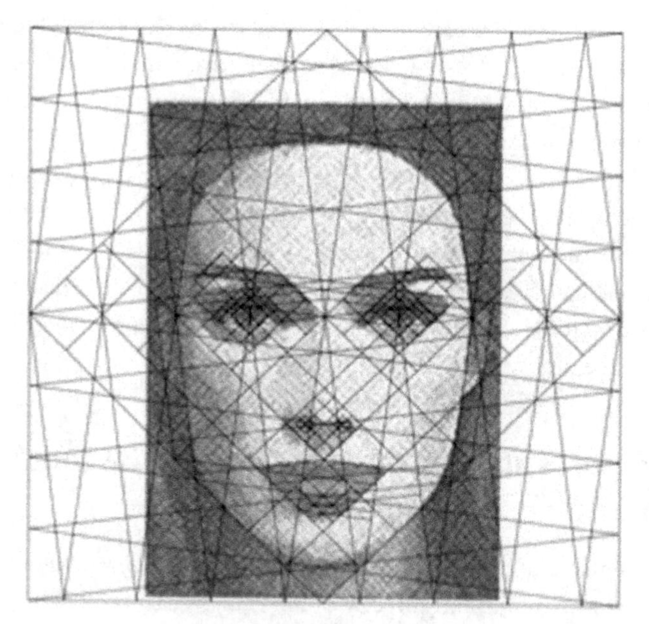

SOURCE: Reproduced with permission. Schmidhuber, J (2009) Simple algorithmic theory of subjective beauty, novelty, surprise, interestingness, attention, curiosity, creativity, art, science, music, jokes, *Journal of SICE*, **48** (1)

There is some suggestive evidence from eye-tracking research that people do sense – even non-consciously – hidden geometry in an image.[22] By keeping the geometry hidden, there is no onus on the viewer to look at it; they are just presented with the more simple surface image. Yet at their own pace their non-conscious mind can explore the hidden pattern behind it. It is similar to the way that Pixar animated movies manage to appeal on two levels: the simple, surface cartoon hijinks of the story, whilst the underlying more sophisticated subjects of jokes and cultural references add depth and interest for parents watching.

Whilst Classical and Renaissance artists and designers may have been motivated to embody geometric patterns in their work, the ability of computers to quickly generate detailed patterns from simple rules provides a potentially wider array of geometric templates. Currently, low-complexity designs based on these templates are hard to produce. But in the future perhaps computers will be able to help with that too, either by suggesting possible designs or images that can be created from a template, or – if given a particular design – modifying it to fit a geometric underlying template, making it more interesting to look at.

Similarly, some designers already use underlying geometric templates to create their designs. For example, the logos for Twitter and for Apple's iCloud system are based on a series of overlapping circles. It is not immediately apparent, but if you look you can see them. Equally, web designers often use grid templates to position elements on the page, creating order and consistency between pages. However, low-complexity design goes further than this. The underlying templates in low-complexity design have their own complex, intriguing geometric patterns that are more involving than a simple grid.

Constructal law

Minimalist design is about finding the simplest, least energy-consuming (in terms of thinking) solution to conveying information or explaining a task. This is something that nature also does.

Devised by Adrian Bejan, a professor of engineering, the *constructal law*[23] claims that any moving or living system – from trees to rivers to our lungs – evolves a pattern or design that enables energy to flow through with the least amount of resistance. It explains how nature creates geometric and structured patterns.

The starting point of the theory is that systems evolve a design because they have to deal with a flow of energy. This could be the flow of water through a landscape (eg a river) or the way that logs and icebergs floating on water become positioned perpendicular to the wind (as this transfers energy from the air to the water more effectively).

The theory is important because it links the design of living things to physics: they are formed by similar processes to each other. This is why we see similar patterns across different types of systems. For example, forked lightning, trees, rivers and lungs all feature similar branching patterns.

▶

The constructal law also points to another feature of minimalistic design: by finding the most efficient way to convey information or allow a user to complete a task, there can be something natural and inevitable about the solution. If it is the best way, it often seems more like discovering a fundamental part of nature than inventing an arbitrary solution (see Figure 3.7). As Apple's Jony Ive describes it: 'So much of what we try to do is get to a point where the solution seems inevitable:... you think "of course it's that way, why would it be any other way?"'[24]

Figure 3.7 Similar basic shapes found in nature are the result of efficient ways to dissipate or deliver energy

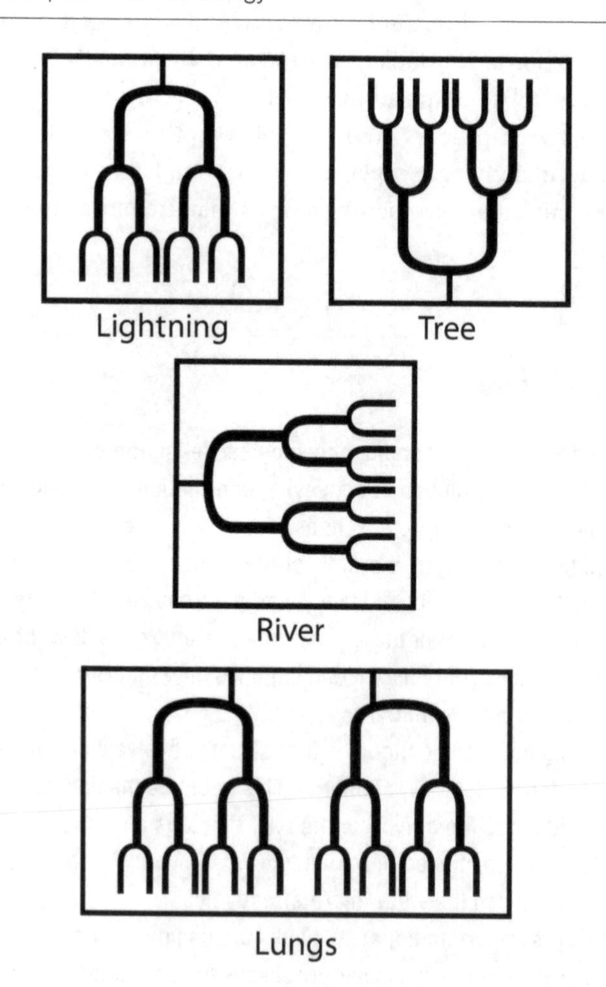

Ways to make designs more fluent

There are a number of techniques for adding processing fluency to a design. Some work by simplifying a design and cutting away redundant visual information, some by tapping into ways that our brains find particular image types easier to process.

Familiar form but intriguing detail

As we have seen, designs that are more complex can be liked if they convey a lot of information per design element (propositional density). Also, simple designs tend to be liked more if they are simpler than the viewer expects them to be.

Neuroscientist Jan Landwehr uses this fact to argue that designs with complex details can be liked if their overall form is familiar.[25] He gives the example of the Opel Corsa car. It has a lot of detail to its design (complexity) yet its overall form and outline are prototypical (simple and familiar). Hence the existence of lots of detail sets an expectation amongst people that it will be more difficult to process, yet its overall familiar easy-to-digest form comes as a surprise and creates a positive viewing experience.

Clarity and contrast

Higher levels of contrast between the subject of an image and the background can increase processing fluency. Research shows that when shapes are viewed with higher levels of contrast to their background, people are more likely to report that they found them attractive.[26]

The effects of clarity/contrast seem to be mainly only if someone is looking at an image quickly. Given longer periods of time (say 10 seconds) the effect disappears.[27]

Self-similar patterns

As we have seen, in low-complexity design, underlying templates of low Kolmogorov-complexity patterns are used as a template for constructing the design. This creates an image that is simple on the surface, but has intriguing compressible information hidden in it. However, these templates may have an additional benefit in making the composition of a design look more naturally integrated and harmonious.

These types of template patterns often exhibit a quality known as self-similarity. The smaller elements of the pattern are similar to the whole. In other words there is a pattern that repeats at different scales. This can create a pleasing inner harmony to the image. Artists and architects often use self-similar patterns as a way to organize their designs. Three examples are the Fibonnacci sequence, fractals and the golden ratio.

The Fibonacci sequence

The Fibonacci sequence (named after the 13th-century Italian mathematician Leonardo Bonacci, who was also known as Leonardo Fibonacci) is a sequence of numbers or shapes in which each new element is based on adding together the previous two elements (Figure 3.8).

Figure 3.8 The Fibonacci sequence

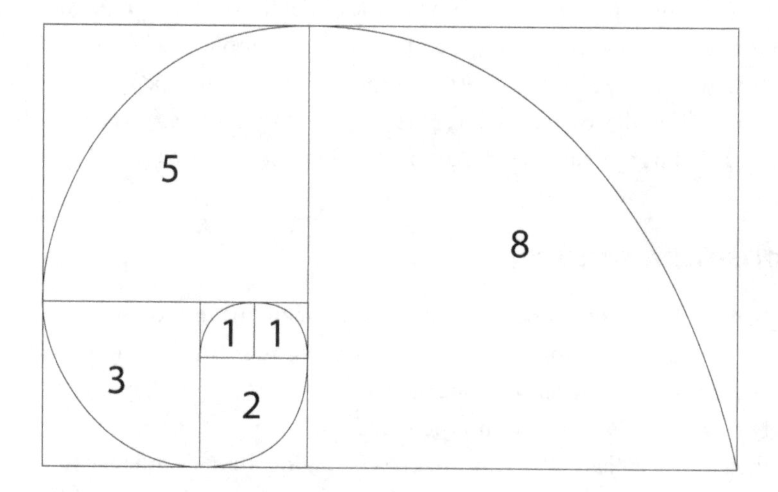

Fractals

The maths on which fractals are based was discovered in the 17th century, but it was not until the advent of cheap computers in the 1970s that it became possible to explore these patterns (see Figure 3.9). The term itself was coined by Polish-born mathematician Benoit Mandelbrot in the mid-1970s. Fractal patterns are everywhere in nature: in clouds, rivers, mountains, coastlines, crystals, snowflakes and even in our DNA.

Figure 3.9 A fractal pattern

The golden ratio

The golden ratio is a pair of lengths in which the difference between the long and short lengths is the same as the difference between the total length and the long length. Figure 3.10 makes it easier to see – if we take two lines, A and B, that are in a golden ratio relationship, and then call their total length C...

Figure 3.10 The golden ratio

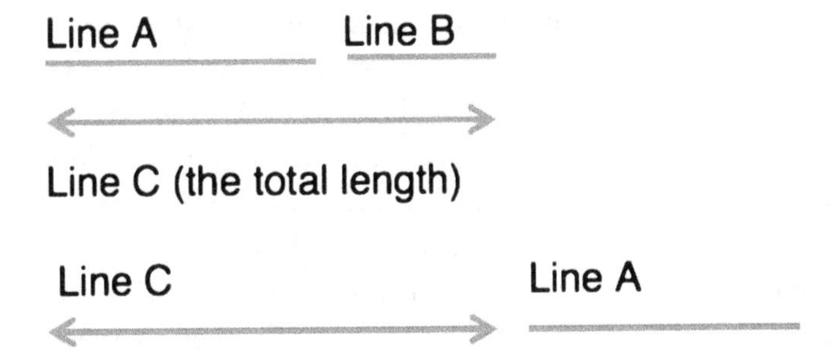

Then put lines C and A together, they will form the same relationship.

If we instead make line A into a square, and line B a rectangle, we get a golden rectangle, with the ratio of height to width being 1:1.618 (Figure 3.11):

Figure 3.11 The golden rectangle

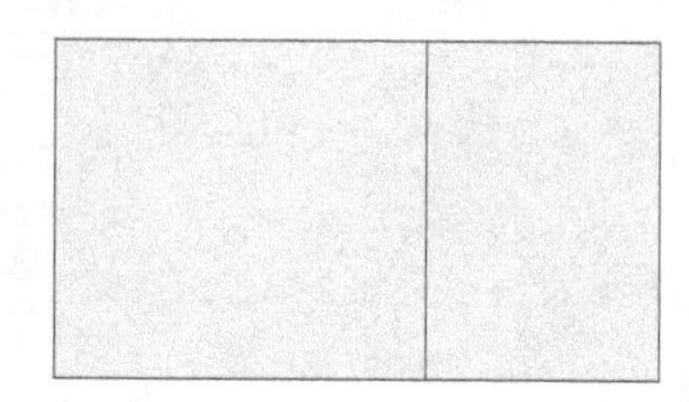

The golden rectangle can be seen in the form of numerous designs of everyday things, including books, doors, credit cards, cigarette boxes, playing cards and televisions (Figure 3.12). The parts (the square and the rectangle) have a relationship to the whole, supposedly creating a pleasing-to-the-eye overall form.

Figure 3.12 Examples of golden rectangle products

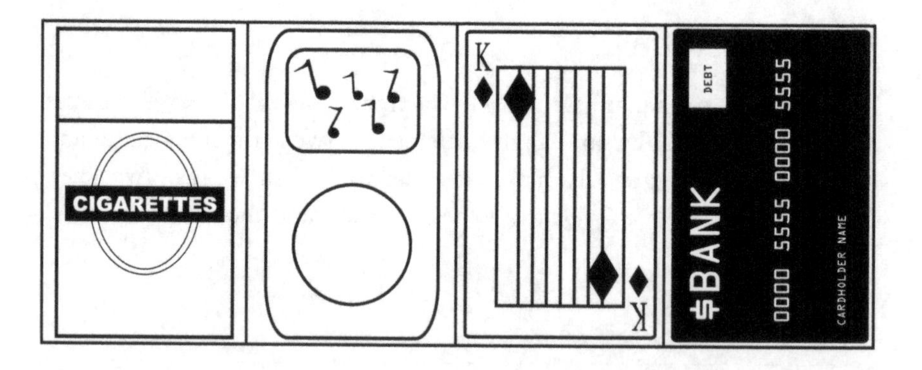

The fact that self-similar patterns are often found in nature may help them feel organic, as well as intentional (ie they create the intuitive feeling that there is a logic to the composition of the design – it is not just random). It could also mean that our brains have evolved an inbuilt ability to detect these patterns (as with the example of Jackson Pollock's fractal paintings, discussed in Chapter 2) as our ancestors would have been constantly exposed to them in their natural environments during our evolutionary history (see Figure 3.13). As well as in the visual arts, self-similar patterns are also used in music and poetry.

Figure 3.13 The Parthenon Temple in Athens exhibits design features that conform to both the golden ratio and the Fibonacci sequence

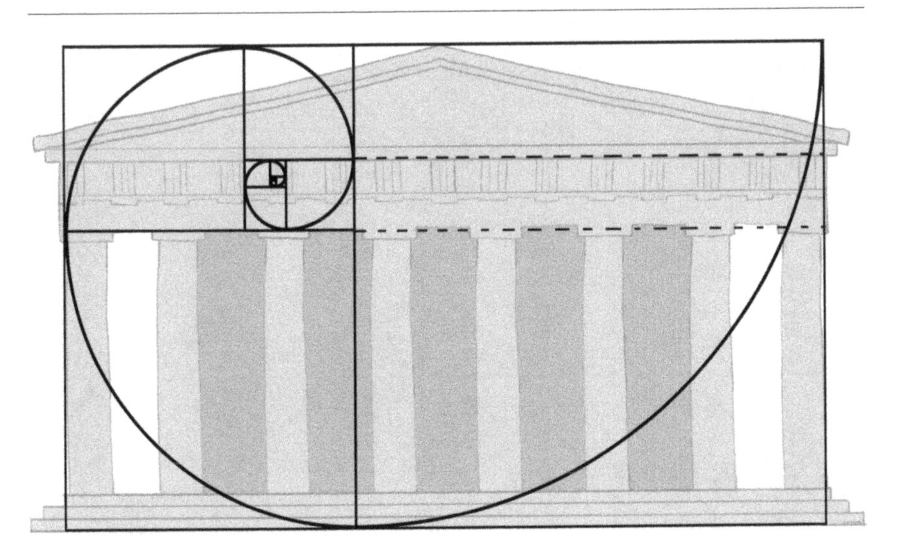

Does the golden ratio 'work'?

The golden ratio and the golden rectangle are widely thought to have been used throughout recorded history as a way to make everything from paintings to buildings more harmonious and beautiful. However, whilst Euclid wrote about the golden ratio back in the 3rd century BC, there is not any direct, firm evidence that Classical-era artists and architects were consciously using it. Although it seems to have been used as far back as the design of Stonehenge, around 5,000 years ago, and in ancient Greece, the idea may have originated in a treatise written in 1854 by a German psychologist called Adolf Zeisig, who described finding the golden ratio in many classical sculptures and architecture.[28]

Experimental evidence that people prefer compositions divided by the golden ratio is mixed.[29] Some studies have found a preference, some haven't. It may be that some people find it beautiful and some don't, thus accounting for the mixed evidence in its favour. Alternatively it may be that there is simply a more general preference for rectangles that are close to the golden rectangle's proportion, rather than it being exact. The preference range in research seems to be rectangles with height to width ratios between 1:1.2 and 1:2.2 (see Figure 3.14). As data visualization expert

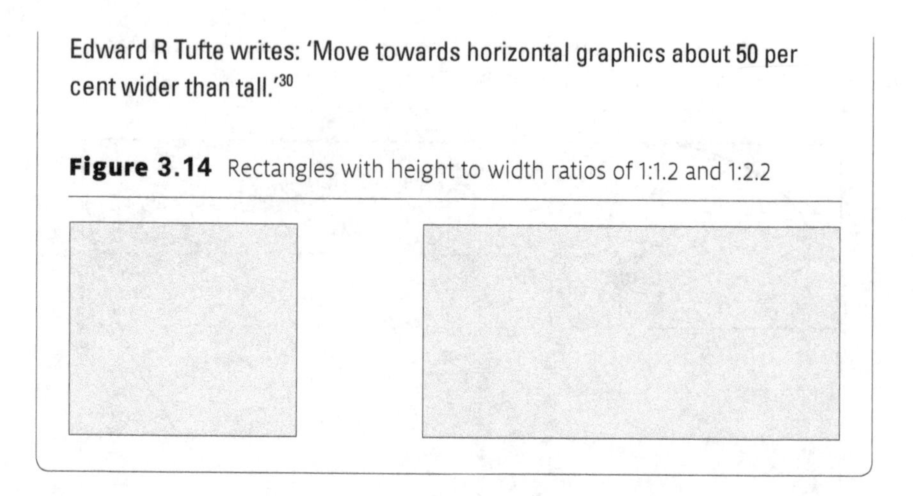

Edward R Tufte writes: 'Move towards horizontal graphics about 50 per cent wider than tall.'[30]

Figure 3.14 Rectangles with height to width ratios of 1:1.2 and 1:2.2

The rule of thirds

A simplified version of the golden ratio that is often recommended to designers and photographers is the rule of thirds: dividing the available space with two equally spaced horizontal and vertical lines, as shown in Figure 3.15:

Figure 3.15 Grid used in the rule of thirds

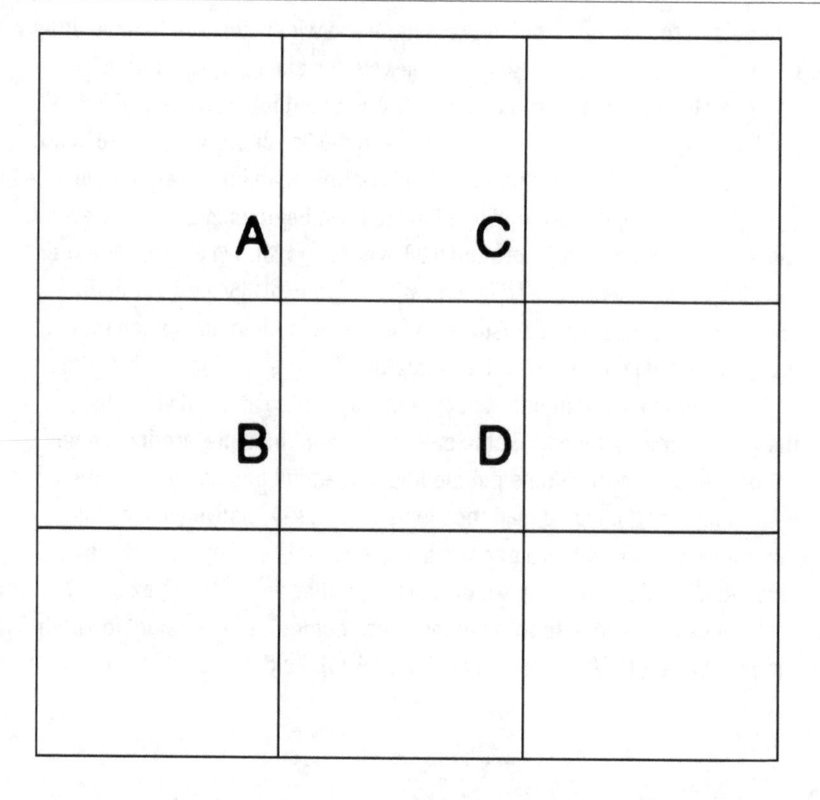

The idea is that, by placing the most important visual information intersecting one of these lines (or, where two of the lines intersect), it will result in a more pleasing and balanced image. Research using landscape photographs showed a significant preference for those based on the rule of thirds; in particular three of the points where two lines intersect (A, B and D) had an effect in people's preference for the images.[31] The greater effect of points on the left visual field may be related to something called the pseudoneglect effect (see below).

Symmetry

Reflectional symmetry is another good example of the regularity of self-similarity – it is a repeating of half the image around an axis of reflection.

Our brains find processing symmetrical images easy. Infants as young as four months old can recognize symmetry, and by 12 months they show a visual preference for it.[32]

Research has shown that people show the highest preference for symmetries around a vertical axis, followed by those around a horizontal axis, with those around a diagonal axis having the weakest preference (Figure 3.16).[33]

Figure 3.16 Symmetries around vertical, horizontal and diagonal axes

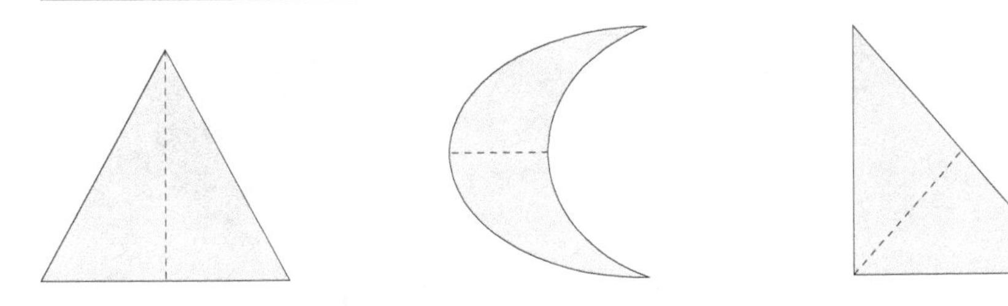

Interestingly whilst the average person tends to prefer simple symmetries, those with art training tend not to.[34] This may be an example of designers not seeing images like the average person does. Their training (or just natural greater levels of interest and perception of aesthetics) enables them to process more complex images more easily. For example, research shows that artists, unsurprisingly, are more skilled at visual perception tasks (such as mentally rotating a 3D-object) than non-artists.[35] Drawing is not merely a mechanical skill of guiding the hand, it develops a better eye for visual understanding. Therefore, symmetry may be an underused technique because designers themselves do not appreciate it.

Left/right differences

There is evidence that designs in which the images are placed to the left, and words to the right, are more pleasing to people.[36] There is a slight processing-fluency advantage to this arrangement. The reason is due to the way our brains process images. As the signals from our eyes are sent to our visual cortex at the back of our brains, the information from the right side of our visual field reaches the left side of our brains first, and the information from the left field reaches the right side of our brains. Our left brain is more specialized at understanding language (this is slightly less likely in left-handed people), whilst our right brain is more specialized at decoding visual patterns (see Figure 3.17).

Figure 3.17 Cerebral dominance and visual field processing

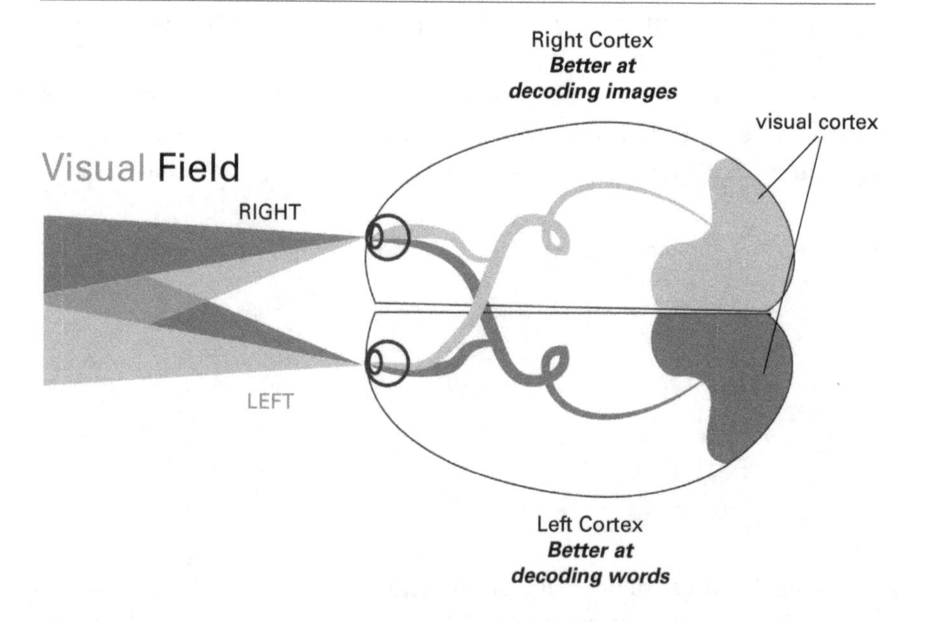

Another interesting effect of this is something called pseudoneglect: our tendency to pay more attention to and be more influenced by visual things going on in our left visual field. For example, take a look at the two bars in Figure 3.18 – which seems overall darker to you?

Figure 3.18 Which bar looks darker?

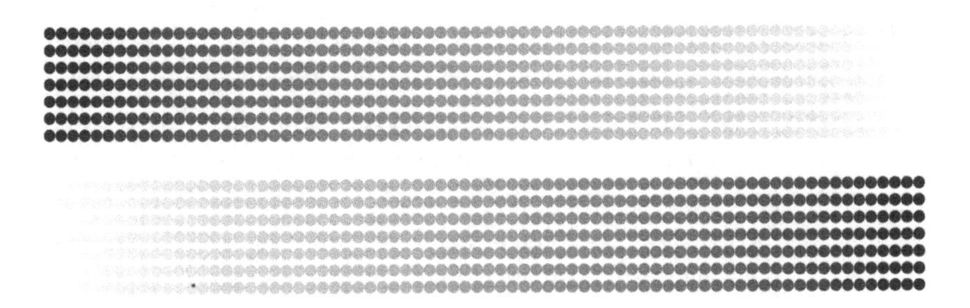

Figure 3.19 Which face looks happier?

In Figure 3.18 most people think the top line is darker, and in Figure 3.19 most think that the left-hand face face is happier, even though in both cases both versions of the images are the same, just mirror images of each other, showing that we overestimate the visual information in the left of each image. When asked to mark the half-way point on a straight line, people tend to place it to the left of its true centre.[37] They are overestimating the amount of the line on the left. People are also faster at detecting changes in their left visual field.[38] Stage-play directors have long known about this effect, and when they need an actor to enter the stage unseen by the audience they will often come in from the right.

These differences are hard-wired into our brains, but culture can also play a role. For example, research has shown that people from cultures that read right to left prefer images that are orientated rightwards, whereas those who read left to right prefer leftwards-orientated images.[39] In the research, French (right-reading) and Hebrew (left-reading) readers were shown a series of images that were orientated to the left or right (such as a vehicle driving rightwards, or a statue pointing leftwards), and mirror images pointing the opposite way. The French readers preferred right-facing images, the Hebrew readers preferred left-facing. Interestingly, there is a bias amongst painters for painting faces turned to the left.[40] However, this could be an example of where one neuro design principle overrides another. By painting the face looking towards the left, it can place the profile of the face more in the left side of the painting, and the back of the head in the right side. As we pay more attention to things in our left visual field it is therefore placing the most interesting part of the head (the face) in the area that we find easiest to decode imagery. Another reason for this is that people seem to prefer looking at the left side of faces.[41] This could also be because we tend to be more emotionally expressive on the left side of our face, hence people expect to gain more useful information about how the person is feeling by looking at this side of their face.[42]

Visual hierarchies

A good visual hierarchy guides viewers' eyes, making it intuitive in what order they should look at things. Similar to the gestalt design principles discussed in Chapter 2, visual hierarchies can be built from our intuitive understanding of how size and position convey importance. For example, larger design elements intuitively feel more important.

Priming and context

A similar effect to the mere exposure effect is that of priming. Context can prime people to find images more fluent. Priming is a psychological effect whereby when we are exposed to something it makes things we strongly associate with it easier to recognize and understand. For example, if you buy a new car or a new coat, you will often suddenly start noticing other people on the street with the same car or coat. It is simply because your awareness has become primed to notice.

In one study, researchers selected a series of images of everyday objects, such as planes, desks and birds.[43] They made different versions of the images that were more or less fluent by degrading the quality of the images (thus making them harder to see). As each image appeared on a computer screen, participants were asked to press a key as soon as they recognized what it was, and then rate how much they liked the image. However, before they had seen the image the researchers had subliminally flashed up a contour outline of either the full image they were about to see, or of a different image. When people had been subliminally exposed to the contour outline of an image, they were then faster in recognizing the image and rated it more favourably. Being subliminally primed had increased the processing fluency of the image.

Priming can be irrational: information from one context can trigger behaviour in a completely unrelated context. For example, during the summer of 1997 sales of Mars chocolate bars increased.[44] During this period the NASA pathfinder robot had landed on Mars and was in the news a lot. Simply hearing the word Mars had primed people to think more about it, and made the very concept of 'Mars' more fluent and mentally accessible. Similarly, research has shown that around Halloween, sales of orange goods increase. The regular sight of pumpkins around this time primes people to think of orange.

The peak shift effect

The peak shift effect, described in Chapter 2, whereby the most distinctive elements of a design are exaggerated to help recognizability (as in caricature cartoons), can also aid processing fluency. Consider what elements in a design are either distinctive or differentiate it from other designs and think about how they could be exaggerated. Can a colour be made bolder or more luminescent? Can a shape be made larger or more exaggerated? A curve curvier or an angle more angular?

Perceptual subitizing

In the movie *Rain Man* (1988) there is a character with autism, played by Dustin Hoffman, based on a real man: Kim Peak. He has some extraordinary abilities, one of which is seen when someone accidentally drops a box of toothpicks, scattering them on the floor, and he almost instantly knows that there are 246, seemingly without having to count them one by one.

Perceptual subitizing is our ability to instantly see how many things there are – such as individual elements in an image – without having to count them one by one.[45] For most of us, unlike Kim Peak, we can only do it for relatively small numbers of things – around three or four. Sometimes we can do more than three or four, such as the patterns of dots on a die. We can instantly recognize the six-face, but these are special cases in which we have already memorized a pattern. It is thought that our brain's ability to do this comes from our hunter-gatherer past. For example, the need to almost instantly size up whether a group of predator or prey animals in the distance are greater or lesser in number than our own small group of hunters: all we would have needed to know is whether there are one, two, three, four or many animals. Knowing the difference between seven or eight prey was not as critical as knowing the difference between one or three.

A design with a minimal number of individual graphical regions that are positioned in an integrated way will be easier for viewers to decode than a design in which there are many elements that are arranged in an unintegrated way. If viewers' eyes have to move around many elements, and maybe even move back and forth, returning to the same element more than once, like a ball ricocheting around a pin-ball machine, it will feel more disfluent.

Similar to perceptual subitizing is the fact that the more 'expert' people are with a particular category, the easier they find it to decode images and designs that relate to it.

Orientation sensitivity and the oblique effect

Analogue clocks are one of the most successful designs of all time, and every day hundreds of millions of people intuitively grasp the time by a glance at the position of the clock or watch hands. However, why have analogue clocks and watches only succeeded in 12-hour format, not 24 hours? The answer may be that our perception of angle differences is sensitive up to about 30 degrees, but smaller differences than that become difficult to decode.[46]

Our visual cortex finds it easier to decode lines that are cardinal – ie vertical or horizontal – than those at an angle: a phenomenon that neuroscientists call the *oblique effect*. Our visual cortex even has groups of neurons specifically sensitive to cardinal lines. Detecting the difference between lines positioned at different angles becomes harder the closer they are together – 30 degrees, as in the different spacings between the numbers on a classic analogue clock, allows us to intuitively tell the time at a glance whereas more numbers on the clock face, with differences smaller than 30 degrees, would not be so intuitive (see Figure 3.20).

Thus in general it may be better to position lines, edges and objects either vertically or horizontally. However, if there are multiple edges at angles, it may be best to keep them at least 30 degrees apart, or exactly aligned. Equally, if you want people to pay more attention to lines – ie force them to work a bit harder to decode what they are seeing – and your image is otherwise very intuitive and easy to grasp, you might decide to place the lines at an angle.

Figure 3.20 12-hour and 24-hour clocks

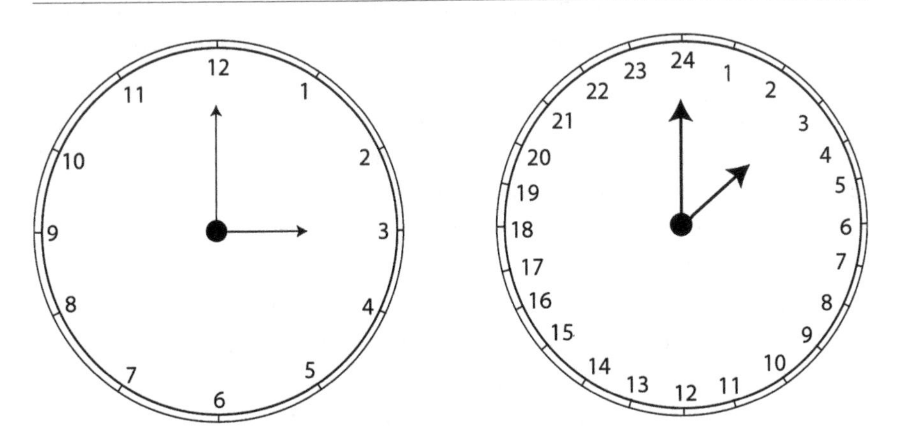

For example, American neuroscientist Stephen Macknik has noticed something curious about the way advertisers and retailers display watches and clocks.[47] Typically they are photographed set to 10:10 (Figure 3.21). This is true across different brands. Macknik traced the practice back to the 1920s and the Hamilton Watch Company. However, the artist Marc Chagall also used it in a series of paintings of clocks, starting in 1914.

Figure 3.21 Watch set to 10:10

Why should they be set to this time?

This time-setting places both hands at an oblique angle, rather than cardinal positions (ie horizontal or vertical). As we know, oblique lines require a tiny bit extra effort to process. Setting a watch to 10:10 is therefore a tiny bit harder to read, and you might think would be less desirable, according to the principles of processing fluency. However, Macknik believes that the tiny bit of extra effort required to read the watch hands set to 10:10 is effective because it forces us to pay extra attention to it. Watches are simple things to look at – by forcing people to pay a little bit more attention they might be more likely to be seen, or seen for longer. (It could also be that watches with hands set to cardinal positions offend our abhorrence of coincidence (see Chapter 2): setting the hands to exactly on the hour, quarter past, or half past looks very deliberate and 'non-typical'. The 10:10 setting is also slightly reminiscent of a smiling mouth.)

Key concepts

Processing fluency

This is how easy and quick it is for people to visually understand an image (perceptual fluency) or its meaning (conceptual fluency). Fluent images create a feeling of familiarity in the viewer.

Mere exposure effect

The more often a person has seen an image, even subliminally, the more they like it.

Propositional density

The number of visual elements in an image divided by the number of meanings it conveys. Density levels above one are good because they convey lots of meaning in a minimal, and hence fluent, way.

Kolmogorov complexity

A measure of an image's complexity: the shortest computer program it would take to generate the image. Patterns with self-similarity (such as the Fibonacci sequence, golden ratio and fractals) tend to have low Kolmogorov complexity.

Low-complexity design

Designs that are based on an underlying pattern with low Kolmogorov complexity.

Embodied cognition and intuitive physics

Embodied cognition is the idea that we use our bodies to help us think. We use feedback from our muscles and senses to tell us things about the world and as shortcuts to make quick assessments. For example:

- If something is heavy, it can seem more important or of higher quality.
- If we have to move something towards us, it can make us feel more positive about it.
- If something is easy we say it feels 'smooth'.
- If we've had a difficult day we say it has been a 'rough' day.

In a sense these are body metaphors. They work quickly and intuitively as we have all built up a rich memory of our experiences through interacting with the world. Embodied cognition metaphors work non-consciously and intuitively, we don't have to think about them. If you need to effortlessly communicate a particular quality – such as importance, happiness, coldness, warmth, etc – consider whether there are bodily actions or sensations that will remind people of this quality, and depict them in images or words. (More on this subject in Chapter 5.)

Showing people, rather than asking them to imagine

Imagining things takes some effort. Some things are harder to imagine than others. Either show people images of what you want them to think about, or paint pictures with words that are concrete and easy to imagine. Sometimes just including an image, such as in a set of instructions, helps lower the cognitive load of a task for users.

Web usability and processing fluency

As we have already seen, web users glance, scan and skim webpages. They don't thoroughly read everything and think through every detail in a considered, in-depth manner. Web usability expert Steve Krug wrote a classic book – *Don't make me think*[48] – on what he believes is the most important rule of web design: minimizing the amount of mental effort that users have to exert in order to navigate a site.

He writes: 'Using a site that doesn't make us think about unimportant things feels effortless, whereas puzzling over things that don't matter to us tends to sap our energy and enthusiasm – and time.'

▶

Krug recommends several key principles for maximizing the fluency of a website:

Designing a page so it can be skimmed

A page should ideally be intuitive enough that it can be navigated quickly and easily, without making users stop to think. As well as some of the techniques outlined in this chapter (eg use of visual hierarchies) Krug advises breaking up a page into clearly defined areas, and making it obvious what is clickable.

Make each choice of click easy

Whilst the number of clicks to perform a task on a site is an often-cited usability metric, Krug believes it is more important to consider the ease of each click. Does the design of the site make it easy, unambiguous and clear where users should be clicking to achieve what they want? Minimizing the mental effort that users have to put into clicking makes using the site feel more effortless.

Minimize the use of words

He recommends getting rid of needless words on a page. In particular, instructions. As web users skim through a site, they generally don't read instructions anyway. Strive to make the design so intuitive that instructions are not needed.

Checking the complexity of your designs

There is no one perfect way to measure the complexity of a design. At a simple level the 'compressibility' of an image when saved on a computer is an approximate measure. In other words, when you save an image on a computer, the computer will compress it. The less information in the image, the smaller the compressed file will be in bytes or megabytes.

However, this is an imperfect measure, as we have already seen that other factors, like symmetry or repeating patterns, can make an image simpler for people to decode. So here are some things to consider when deciding how simple your image is:

1. Does the image have any symmetries?
2. Does it conform to any underlying template pattern?
3. What is its propositional density?
4. How many main individual elements does it have?
5. Is it easy for people to take in as they look at it, or will their eyes jump around trying to decode the different elements?
6. Is there a natural hierarchy of importance to help guide people where to look?
7. Does the design have any unnecessary elements that could be removed?

Summary

- Our brains have a bias for imagery that they find easy to process. In particular, we like information that turns out to be easier to process than we expected.
- When images are easy for us to process, we tend to feel more positive about them, and the converse when they are harder to process. This can give an advantage to simpler designs.
- We have a slight bias towards images we have seen before – even if they have been presented so fast we are not consciously aware of seeing them. This is called the *mere exposure effect*.
- Another form of familiarity is prototypicality, or the *beauty in averageness* effect. People have a bias towards images such as faces and cars that look average.
- More complex images can be preferred over more simplistic images if they carry a lot of meaning. This level of meaning is called *propositional density*.
- When images hold hidden structure, or the promise of teaching us a pattern we didn't know before, they can become more interesting.
- Some of the effects of things like clarity and simplicity are stronger over short durations and weaker, or reversed, on repeated exposures as they become more familiar.

Notes

1 Jodard, P (1992) *Raymond Loewy*, Trefoil publications, London.

2 Dieter Rams, quoted in: http://www.theguardian.com/artanddesign/artblog/2008/jan/16/applebrauniverams (last accessed 25 August 2016).

3 http://theoregonsampsons.blogspot.co.uk/2012/08/an-incident-from-academia-black-bag.html (last accessed 25 August 2016).

4 Zajonc, RB (1968) Attitudinal effects of mere exposure, *Journal of Personality and Social Psychology*, **9** (2, Pt 2), p 1.

5 Bornstein, RF (1989) Exposure and affect: overview and meta-analysis of research, 1968–1987, *Psychological Bulletin*, **106** (2), p 265.

6 Topolinski, S, Likowski, KU, Weyers, P and Strack, F (2009) The face of fluency: semantic coherence automatically elicits a specific pattern of facial muscle reactions, *Cognition and Emotion*, **23** (2), pp 260–71.

7 Bolte, A, Goschke, T and Kuhl, J (2003) Emotion and intuition effects of positive and negative mood on implicit judgments of semantic coherence, *Psychological Science*, **14** (5), pp 416–21.

8 Harmon-Jones, E and Allen, JJ (2001) The role of affect in the mere exposure effect: evidence from psychophysiological and individual differences approaches, *Personality and Social Psychology Bulletin*, **27** (7), pp 889–98.

9 Winkielman, P, Halberstadt, J, Fazendeiro, T and Catty, S (2006) Prototypes are attractive because they are easy on the mind, *Psychological Science*, **17** (9), pp 799–806.

10 Unkelbach, C and Greifeneder, R (2013) *The Experience of Thinking*, Psychology Press, London.

11 https://techcrunch.com/2013/06/11/jony-ives-debutes-ios-7-bringing-order-to-complexity/ (last accessed 9 October 2016).

12 Graf, LK and Landwehr, JR (2015) A dual-process perspective on fluency-based aesthetics: the pleasure–interest model of aesthetic liking, *Personality and Social Psychology Review*, **19** (4), pp 395–410.

13 Carbon, CC, Faerber, SJ, Gerger, G, Forster, M and Leder, H (2013) Innovation is appreciated when we feel safe: on the situational dependence of the appreciation of innovation, *International Journal of Design*, 7 (2), pp 43–51.

14 Landwehr, JR, Wentzel, D and Herrmann, A (2013) Product design for the long run: consumer responses to typical and atypical designs at different stages of exposure, *Journal of Marketing*, 77 (5), pp 92–107.

15 Martindale, C, Moore, K and Borkum, J (1990) Aesthetic preference: anomalous findings for Berlyne's psychobiological theory, *The American Journal of Psychology*, **103** (01) pp 53–80.16

16 Tufte, ER and Graves-Morris, PR (1983) The visual display of quantitative information, vol. 2, no. 9, Graphics Press, Cheshire, CT.

17 Dechêne, A, Stahl, C, Hansen, J and Wänke, M (2009) Mix me a list: context moderates the truth effect and the mere exposure effect, *Journal of Experimental Social Psychology*, **45** (5), pp 1117–22.

18 Schmidhuber, J (2009) Simple algorithmic theory of subjective beauty, novelty, surprise, interestingness, attention, curiosity, creativity, art, science, music, jokes, *Journal of SICE*, **48** (1), pp 21–32.

19 Halberstadt, J and Rhodes, G (2003) It's not just average faces that are attractive: computer-manipulated averageness makes birds, fish, and automobiles attractive, *Psychonomic Bulletin & Review*, **10** (1), pp 149–56. (Note: as well as average or prototypical images appearing more familiar, there may also be a more direct evolutionary effect of averageness: there is some evidence that natural forms – such as faces – contain more genetic 'fitness' if they appear average.)

20 http://people.idsia.ch/~juergen/locoart/locoart.html (last accessed 25 August 2016).

21 The original reference to low-complexity art in general is: Schmidhuber, J (1997) Low-complexity art, *Leonardo*, pp 97–103.

22 Bartlett, C (2011) The eyes have it: focal point choices and compositional geometry in painting, *Proceedings of Bridges 2011: Mathematics, music, art, architecture, culture*, pp. 489–92, Tessellations Publishing.

23 Bejan, A and Lorente, S (2010) The constructal law of design and evolution in nature, *Philosophical Transactions of the Royal Society of London B: Biological Sciences*, **365** (1545), pp 1335–47.

24 http://www.macworld.co.uk/news/apple/13-most-philosophical-jony-ive-quotes-3490442/ (last accessed 25 August 2016).

25 Landwehr, JR (2015) Processing fluency of product design, in Batra, R, Seifert, C and Brei, D (eds), *The Psychology of Design: Creating consumer appeal*, Routledge, New York, p 218.

26 Reber, R, Schwarz, N and Winkielman, P (2004) Processing fluency and aesthetic pleasure: is beauty in the perceiver's processing experience? *Personality and Social Psychology Review*, 8 (4), pp 364–82.

27 Reber, R and Schwarz, N (2002) The hot fringes of consciousness: perceptual fluency and affect, *Consciousness & emotion*, 2 (2), pp 223–31.

28 Gamwell, L (2016) *Mathematics and Art: A cultural history*, Princeton University Press, Oxford.

29 McManus, IC, Cook, R and Hunt, A (2010) Beyond the golden section and normative aesthetics: why do individuals differ so much in their aesthetic preferences for rectangles?, *Psychology of Aesthetics, Creativity, and the Arts*, **4** (2), p 113.

30 Tufte, ER and Graves-Morris, PR (1983) *The Visual Display of Quantitative Information*, vol. 2, no. 9, Graphics Press, Cheshire, CT, p 190.

31 Svobodova, K, Sklenicka, P, Molnarova, K and Vojar, J (2014) Does the composition of landscape photographs affect visual preferences? The rule of the golden section and the position of the horizon, *Journal of Environmental Psychology*, **38**, pp 143–52.

32 Bornstein, MH, Ferdinandsen, K and Gross, CG (1981) Perception of symmetry in infancy, *Developmental Psychology*, **17** (1), p 82.

33 Palmer, SE (1991) Goodness, gestalt, groups, and Garner: local symmetry subgroups as a theory of figural goodness, in Lockhead, GR and Pomerantz, JR (eds) *The Perception of Structure: Essays in honor of Wendell R. Garner*, pp 23–39, American Psychological Association, Washington, DC.

34 McWhinnie, HJ (1968) A review of research on aesthetic measure, *Acta Psychologica*, **28**, pp 363–75.

35 Kozbelt, A (2001) Artists as experts in visual cognition, *Visual Cognition*, 8 (6), pp 705–23.

36 Ellis, AW and Miller, D (1981) Left and wrong in adverts: neuropsychological correlates of aesthetic preference, *British Journal of Psychology*, **72** (2), pp 225–29.

37 Jewell, G and McCourt, ME (2000) Pseudoneglect: a review and meta-analysis of performance factors in line bisection tasks, *Neuropsychologia*, **38** (1), pp 93–110.

38 Iyilikci, O, Becker, C, Güntürkün, O and Amado, S (2010) Visual processing asymmetries in change detection, *Perception*, **39** (6), pp 761–69.

39 Chokron, S and De Agostini, M (2000) Reading habits influence aesthetic preference, *Cognitive Brain Research*, **10** (1), pp 45–49.

40 McManus, IC and Humphrey, NK (1973) Turning the left cheek, *Nature*, **243**, pp 271–72.

41 Blackburn, K and Schirillo, J (2012) Emotive hemispheric differences measured in real-life portraits using pupil diameter and subjective aesthetic preferences, *Experimental Brain Research*, **219** (4), pp 447–55.

42 Reber, R, Winkielman, P and Schwarz, N (1998) Effects of perceptual fluency on affective judgments, *Psychological Science*, **9** (1), pp 45–48.

43 Berger, J and Fitzsimons, G (2008) Dogs on the street, pumas on your feet: how cues in the environment influence product evaluation and choice, *Journal of Marketing Research*, **45** (1), pp 1–14.

44 Berger, J (2013) *Contagious: Why things catch on*, Simon and Schuster, New York, p 70.

45 Clements, DH (1999) Subitizing: what is it? Why teach it?, *Teaching Children Mathematics*, **5** (7), p 400.

46 Lidwell, W, Holden, K and Butler, J (2010) *Universal Principles of Design, Revised and Updated: 125 ways to enhance usability, influence perception, increase appeal, make better design decisions, and teach through design*, Rockport Publications, London.

47 Macknik, SL, Di Stasi, LL and Martinez-Conde, S (2013) Perfectly timed advertising, *Scientific American Mind*, 24 (2), pp 23–25.

48 Krug, S (2005) *Don't Make Me Think: A common sense approach to web usability*, Pearson Education, India, p 19.

How first impressions work

04

Figure 4.1 Our first impression of a design shapes our eventual opinion of it. Like a fractal, our reactions become more complex over time, but they are just elaborations of that initial first glance

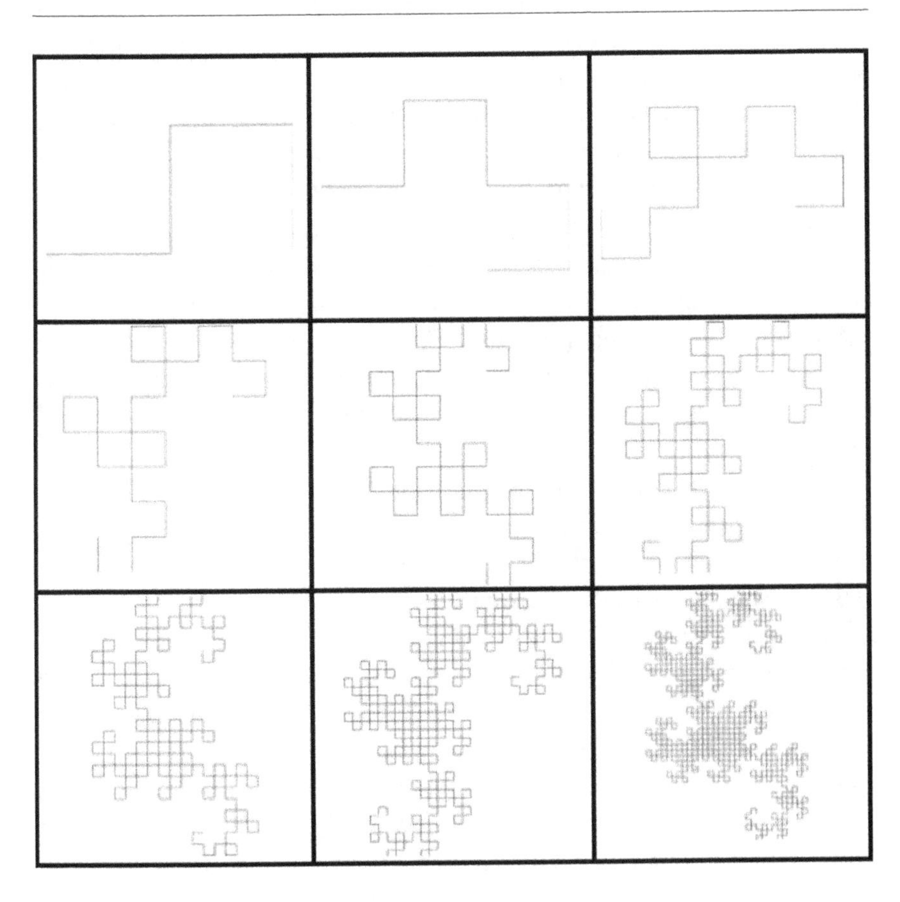

Imagine that you are walking down the street and a new model of car happens to drive past you on the opposite side of the road. It passes your field of view but you are not paying any attention to it as you are thinking of other things at the time. Perhaps you are not even that interested in cars. However, is it possible that even without paying attention to it, your brain has already formed an opinion on what it feels about that car?

Thanks to an experiment conducted by a group of neuroscientists in Berlin we now know the answer to this, perhaps slightly unusual, question.[1] They showed people some images of cars, each for only 2.4 seconds, whilst their brains were being monitored as they lay in an fMRI scanner. For half the group, they were told to look at the cars and rate how attractive they thought each one was. The other half of the group were exposed to the same cars, but whilst they were performing a task requiring them to fixate their eyes at a point away from each car, and they were not asked to rate the cars. Afterwards both groups were asked to imagine they were now faced with the decision of buying a new car; they were shown each of the previous car photos and asked for each one whether or not they would want to buy it.

When the researchers came to analyse the brain data of the participants, they found that by looking at their brain activity at the time the car images appeared on the screen they were able to predict whether they would subsequently want to buy the car or not, no matter whether they had actively looked at the car or whether it had merely appeared within their field of view whilst they were looking at something else. In both cases the brain activity was equally predictive. This shows that our brain can make automatic decisions about whether or not we would want to buy something, even if we see it both very quickly and don't even actively look at it and think about it.

This feels a bit counter-intuitive – almost like our brains are making up our minds without even consulting us!

We are all familiar with our natural tendency to appraise people at a glance, despite knowing that it is often unfair and irrational. We can instantly get a feel for a person: whether we think they are friendly, intelligent, trustworthy and so on. For example, one study showed that participants were able to accurately predict the effectiveness of a salesperson (as rated by their manager) after just hearing a 20-second audio clip of them.[2] Also, interviewers' ratings of a job candidate are the same within two seconds of seeing them as after interviewing them.[3] We are also surprisingly quick at grasping the gist of a picture. Research shows, for example, that people can tell within one-tenth of a second whether an image is of something natural or man-made, or whether it is showing an indoors or outdoors scene.[4]

Such quick assessments probably have an evolutionary origin: back in our hunter-gatherer past it would have been essential to quickly 'size-up' a stranger to assess whether they might be friend or foe. Such judgements back then could have made the difference between life and death. Evolution just didn't equip us with the ability to always hold back and seek more information before we evaluate something: we seem to be primed to create an almost automatic, reflexive judgement on things. The characteristics that we use to size-up people on our first impression are hence tailored to these evolutionary pressures – trustworthiness and attractiveness, for example, to assess how threatening someone might be, or how genetically viable they might be as a potential mate. However, what most of us are less aware of is the extent to which we may be doing such quick judgements all the time. Not only for faces, but for webpages, designs and ads.

When a webpage is loading and its images, logos and text are beginning to appear, the sensible, rational approach might be to reserve judgement until we have had a chance to properly study the page, read it and weigh up its content. Yet instead we mostly form a snap judgement based on its overall look and feel, in which surprisingly few elements of the page have a disproportionate effect on how we subsequently rate it; it can also happen very fast. Experiments have shown that people decide whether they like a webpage within 0.05 seconds of seeing it.[5]

Whilst webpages differ widely on their content, their products, corporate personality, ethos and many other factors, the fact that users are making such fast judgements means that they are not really considering the content at all, they are merely reacting to the overall gist of the design.

The halo effect

This initial snap judgement would not necessarily be that important for designers if it weren't for another interesting fact about first impressions – they tend to last. Psychologists already know about this phenomenon and call it the 'halo effect' or the tendency for a positive feeling about something to subsequently non-consciously bias us to rate it positively on other factors too. We are more likely to bring our conscious judgements of the page into line with this first-impression gut feeling: if the initial emotional judgement is positive, we will then find reasons to assign positive qualities to the page.

First impressions arrive as feelings, which then may be rationalized. Our emotions can be triggered extremely rapidly, before we have a chance to consciously decode what we are seeing. For example, some research has shown that emotional expressions can begin to appear on people's faces as soon as a few milliseconds after seeing an image (Figure 4.2).[6]

For this reason, if you rely on just asking users of your page why they like or dislike it, you could arrive at misleading conclusions. The user is probably unaware that they are making such snap, intuitive judgements and are likely to instead think of rational reasons why they like or dislike the page.

A specific example of the halo effect is the so-called 'beautiful is good' effect: this is our tendency to imbue attractive people with all sorts of desirable qualities that have nothing to do with their levels of physical attractiveness at all. For example, we may think they are more intelligent, trustworthy and reliable.

Figure 4.2 The process of first impressions guiding our view of a design

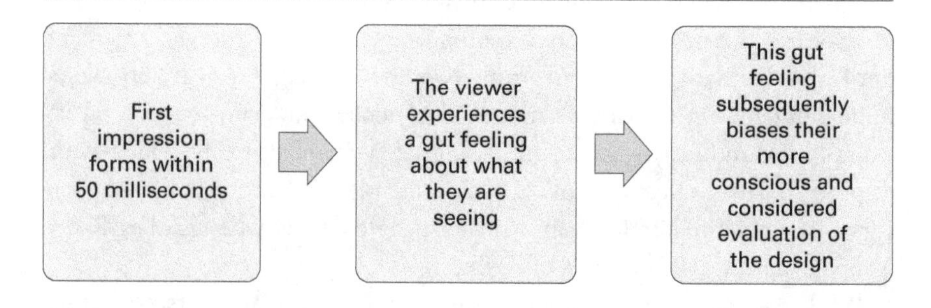

Are first impressions just a feeling?

Neuroscientists call this quality of the first impression of an image 'visceral beauty'.[7] This is the spontaneous gut reaction to the general look and feel of a design. Some experts have questioned whether first impressions are genuinely an evaluation of beauty. They argue that beauty is an idea that needs some kind of thought process to judge, and there just is not enough time to do that in a first impression. Under this model, all that happens is that the first impressions create either a positive or negative feeling, which may then bias our later perceptions of things like trust, usability, attractiveness and novelty. In other words, it's a simple halo effect.

However, there is some evidence to suggest that it is more than just the halo effect and that our first impression also includes a deeper appraisal of the webpage. One study found evidence that both attractiveness and novelty ratings of a page are determined at our first impression, with weaker evidence that usability and trust are also determined at first glance.[8] There is also some evidence that people will quickly reject a site if they find it unattractive.[9]

Whilst the evidence for first impressions of people's ratings of a site's trustworthiness and usability is more mixed, the evidence still shows that these things are driven by a site's design. To judge how much we should trust a website (assuming it is not a well-known site), if we were being logical we should investigate its reputation, read the 'small print' (eg the privacy policy) etc. Yet this is too time consuming and impractical for most of us most of the time so instead we seek a shortcut to inferring it. Some research shows that a design feature they call 'craftmanship' is a good predictor of how users rate a site's trustworthiness.[10] Craftmanship is the appearance that the site has been designed with skill and care, using the latest techniques and technologies. Equally, there is evidence that design aesthetics do influence user judgements of usability. Visceral beauty in a website can even trump actual usability: if a page has poor usability, users will still like it if they find it attractive.[11]

Therefore it may be that some of the important aspects of a website get evaluated in the first impression, whilst others are biased by it but not completely determined instantly, yet they are still also influenced by the design as the user sees more.

First impressions of people online

People like the ability to quickly form an impression of someone. It reduces our uncertainty about them, making it easier to know how to interact with them. Professor of Social Psychology Frank Bermoeri describes an effect called the 'expressivity halo': 'people who communicate in an expressive, animated fashion tend to be liked more than difficult-to-read people; even if they are expressing something such as irritation. Because we're more confident in our reading of them, they are less of a threat.'[12] In other words, we like to be able to get a good, accurate first impression of people, it helps us to know how to interact with them.

First impressions of people have been the subject of many business and personal development books, and people are becoming more aware of the notion of 'personal branding' in the way they present themselves professionally. There are often times when web users will be getting first impressions of people online, whether it is profiles on sites such as LinkedIn or Facebook, or just on the 'about us' page of a website, where users will be judging the people behind your company. Often your online profile photos, be they on your website or a social media site, are the first impression that potential clients or hirers get of you.

First impressions in person form quickly and seem to be long lasting, but can take in a rich array of information: the person's height, eye contact, posture, tone of voice and so on. Online, however, a person's photo carries less dynamic information, yet some research shows that users of social media sites can still make quick inferences about people in profiles even after viewing minimal details. Moreover, these first impressions are predictive of their subsequent evaluations after more fully exploring their profiles.[13] Perhaps unsurprisingly, the profile photo was one of the most frequently used bits of information that people used to make their fast judgements.

Such quick assessments place a disproportionate amount of importance on our photos and mean that even traits that are nothing to do with our face still get judged on it. For example, research has shown that on LinkedIn, men with beards were viewed more favourably for a job requiring expertise (and hence would have been more likely to be interviewed).[14] The same was true for women wearing spectacles versus when they weren't wearing them.[15] As employers increasingly use such online profiles to research candidates, these effects can make a real impact on people's careers.

It is difficult to make universal recommendations about photos, as there are effects of context. For example, the type and style of photo that may work best for someone on a dating website might not work as well on a career site. Equally there are cultural effects. There can be more differences in rating of first impressions of different photos (angles, lighting, facial expression etc) of the same person than of photos of different people. All this means that it is probably worth testing several different photos and getting quick reactions from people. There are websites where you can test out different face photos (such as www.photofeeler.com) and get people to rate them for different qualities that you are interested in projecting.

The effect on browsing behaviour

The first-impression effect is particularly relevant to websites, given that users tend to often spend just a few seconds on a webpage before deciding whether to click away. Whether a website evokes a good or bad first impression could be critical to triggering the web user's decision to stay on the site or leave.

For example, the largest platform for online video ads, YouTube, uses a system called TrueView, which gives viewers the option to skip the ad after five seconds. Even other online video ad formats are likely to create impatience if the viewer really doesn't want to watch the ad after the first few seconds. Equally, the reach and popularity of YouTube give users an expectation of being able to skip an ad after five seconds if they are not interested. Either way, all online video ads are effectively 'skippable'.

Google have conducted their own research on what types of ads are most likely to be skipped after the first crucial five seconds.[16] They found:

- There is a genuine tension between showing the brand too early or too late: the former risks turning viewers off, the latter risks leaving them without any brand memory. Their recommendation is to show the brand as part of the product, not as a floating logo.
- Humour works well. Perhaps unsurprising, as a humorous ad delivers benefits to the viewer (making them smile, laugh or feel good) regardless of whether they are interested in learning more about the brand or product.
- If not humour then at least some emotional tone, particularly something suspenseful, in the first five seconds works well.
- Again, perhaps unsurprisingly, a recognizable face within the first five seconds helps to avoid viewers clicking away.

First impressions are mainly about engaging emotions. They are usually too fast to be about conscious or rational evaluation of content. If video content does not quickly engage us emotionally, we click away.

Thin slicing and the impatient consumer

Even if the first-impression effect did not exist, the importance of evoking emotions and concepts quickly is becoming increasingly important. As consumers we now typically have very short attention spans and are incredibly impatient. For many purchase decisions we don't invest too much time or consideration. We just get a quick feel for whether we like a brand or product and whether we think it matches what we are looking for. Psychologists call this type of conclusion drawn on small bits of evidence 'thin slicing'.

As we have seen in Chapter 1, the web particularly encourages this type of impatient browsing. There are so many choices available to us online and little effort involved in clicking from one site to another that we are more impatient and faster when shopping online than in real shops. Also we become really good at judging webpages quickly as we have seen so many of them, and have a lot of experience with the types of pages we have used before and what our experiences have been like with them.

These types of quick judgements that web users make on the basis of their existing knowledge of webpages are a little like the quick assessments that experts are able to make. For example, antiques or painting experts are often able to almost instantly size-up a piece of furniture or art and tell if it is fake or not. They cannot always say how they know, but they are accessing a non-conscious store of memories and knowledge. Similarly, web users learn a lot about webpages and can learn to associate different types of designs with different sites, even if they are not able to consciously describe how they are doing it.

Don't make me wait!

Most webpages take between two to three seconds to load.[17] Web connection speeds may be getting faster, but the pages themselves are getting more complex – including more graphics, animations and videos. It is an ongoing battle between connection speeds and volume of content. Two to three seconds may sound fast, but it can feel slow. How can web designers create content-rich pages that don't lose impatient visitors who might click away before their page has even finished loading?

The secret is distraction. The evidence shows that if web users are given some type of animation to look at whilst the page is loading, it keeps their attention busy and avoids them impatiently clicking away. Both Google and Facebook, for example, use animated placeholders. These are lines and boxes that show where the text and images will be loading, with an animated shadow cycling over them, the fast movement non-consciously conveying that things are loading quickly in the background.

Research shows that the design of progress bars can affect users' perception of time, making them feel like they are not waiting too long. The most effective design was one that used a 'ribbing' effect. A normal progress bar shows just a uniformly coloured block moving from left to right, gradually approaching the end of the bar. The ribbed version adds a series of vertical lines – or ribs – into the pattern of the bar (see Figure 4.3). The most effective version of all was where the ribbing itself was moving in the opposite direction to the progress of the bar itself, tricking users' brains into feeling the bar was moving faster than it actually was.[18]

Figure 4.3 Regular (top) and ribbed (below) progress bars

What drives first impressions?

As the first-impression effect occurs so quickly, neuroscience can provide clues as to what might be driving it. Neuroscientists know how long it takes for various visual features to be perceived and then understood. The time frame within which first impressions operate means it is likely that only what are called low-level visual features are being perceived. It is not being driven by the detail of the content, and certainly not by the meaning of the words (there wouldn't be time to read anything).

Katharina Reinecke, a neuroscientist at Harvard University, along with her colleagues, has conducted a number of studies on what is driving users' first impressions to websites.[19] Based on prior research that had shown image complexity and colour to be two of the most prominent, noticeable image features that are visible at first-impressions speed, they decided to focus on these. Another benefit of these features is that they were able to generate computer models that can analyse an image and quantify these features in a way that appears to correlate with how people rate their complexity and colourfulness. In other words, these features can be objectively measured and given numerical ratings. Colourfulness was made up mainly of the images' hues, whereas complexity was essentially the amount of different images and text as well as the variety of colours contained on the page (they also tried using the overall image attributes of balance, symmetry and equilibrium – concepts important in gestalt psychology, as described in Chapter 2 – but whilst they were found to have a statistically significant effect, their impact on image complexity was weak, so they were discarded). In the studies, 242 participants then viewed 450 website images and rated them for visual appeal. The levels of complexity and colourfulness alone accounted for about half of the variation in ratings. In order to get this level of prediction they also had to take into account some demographic variables. It turned out that education level, for example, had a significant effect on how much colourfulness a person liked, whilst age had a significant effect on their preferred level of complexity. The optimal levels of complexity and colourfulness can be plotted like an inverted U on a graph (see Figure 4.4). The inverted Us are slightly skewed, as the dislike of high complexity and colourfulness was stronger than low levels. In other words: it is riskier to err on the side of high complexity or high colourfulness than low.

Figure 4.4 Graphs showing preference for different levels of complexity and colourfulness (based on Reinecke and Gajos, 2014)

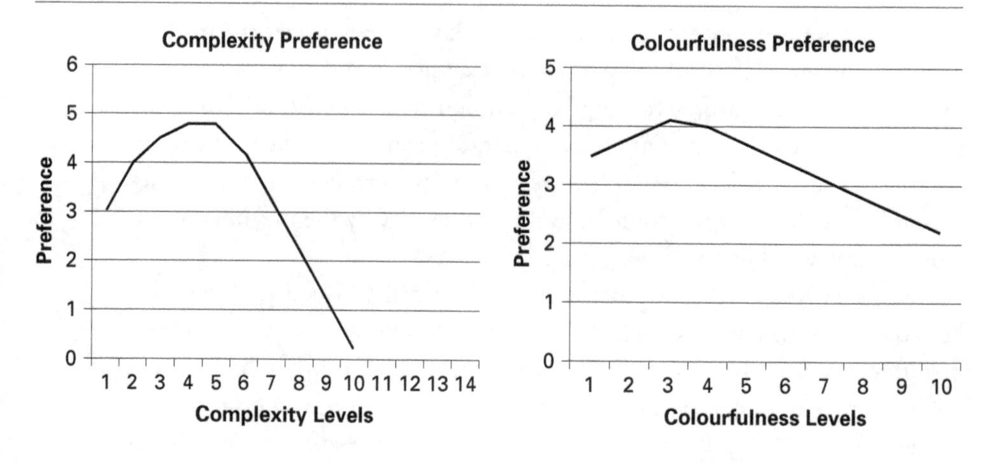

A follow-up study looked more closely at the effect of culture.[20] In the study, 2.4 million ratings of webpages were gathered from across 40,000 people, representing a range of ages, nationalities, education levels and gender. Again, the results showed the influence of visual complexity and colourfulness on first impressions. Participants, overall, preferred a low to moderate level of complexity with a good balance between text and images, and a moderate to high level of colourfulness.

However, there were significant differences between people according to their demographics. For example, those in their thirties preferred a slightly less colourful design than those younger or older (although it is not fully clear yet if this is due to their age, or the tastes of their generations). Interestingly, whilst participants from teenage years to 40 preferred about the same level of visual complexity, older participants preferred a higher level of complexity. This is perhaps slightly unexpected, as you might expect older people to prefer a simpler design that would place fewer demands on their cognitive load. Again, however, it could be something to do with generational tastes, or perhaps with the way that older people are used to interacting with sites (for example, they may be used to giving a website their full attention, whilst younger people are typically more used to doing other things at the same time). In general, women preferred more colourful websites than did men, but men preferred sites that used contrasting colours that helped differentiate the various elements on the page. The optimal level of complexity was about the same for both genders (although women liked low-complexity sites less than did men). Many cultural differences were measured. Whilst the general 'inverted U' pattern (with a higher right side) held true across countries, the peak of the curve differed. For example, Russians tended to prefer simpler designs than did Mexicans; French and Germans preferred less colourful sites than did British or American users. Countries geographically close to each other had similar preferences. For example, northern Europeans generally preferred less colourfulness than did southern Europeans. Education levels altered preferred colourfulness more than preferred complexity: the higher the level of educational attainment, the lower the preferred colourfulness.

Whilst there is an optimal level of complexity for a site to achieve a good first impression, the risks of making a site too complex are greater than making it too simple. In reality most home pages will need to have a minimum level of complexity to them anyway, so most do not risk bad impressions through oversimplicity.

The importance of demographic differences is striking, given that people had less than half a second to view the sites. Whilst there is an overall recommendation here (moderate complexity and colourfulness are best, but lower levels of complexity are less risky than higher), these results also highlight the importance of thinking about who your audience are – doing actual testing or research amongst your users is important.

Some research has also found that 'prototypicality' affects first impressions.[21] Prototypicality (as you may remember from Chapter 3) is how close something looks to how you would expect it to look. Webpages have certain conventions that evolve over time and people get used to them: for example, where users expect the main logo, menu, search bar etc, as well as the general amount of images, text and links they expect to see on a page of that type. An atypical page could create uncertainty: have I arrived at the page I expected? Why does it look different to what I expect? Am I going to have to learn a completely new type of page structure in order to understand what I am seeing?

'Users spend most of their time on **other** sites,' explains web researcher Jacob Nielsen. 'Thus, anything that is a convention and used on the majority of other sites will be burned into the users' brains and you can only deviate from it on pain of major usability problems.'[22]

Of course, it is also possible that prototypicality is being influenced by other design factors: through competition, trial and error and previous testing, sites will tend to cluster around optimal design formats, and then these formats become known and expected by web users.

Interestingly, users seem to judge complexity faster than prototypicality, or at least the effects of prototypicality take slightly longer to make themselves felt.[23] This is probably because prototypicality is a more complicated construct than mere simplicity/complexity of a design. We can easily tell whether a page looks cluttered or simple, but it takes a little longer to compare a page's design to our memory/expectation of what a page of that type should look like.

The optimal combination seems to be low complexity and high prototypicality. Both factors, however, need to be strong: an atypical design cannot be saved by low complexity, and a complex design cannot be saved by high prototypicality.

First impressions may also work through priming. The associations that the first view of something triggers can activate certain concepts and emotions that we then subsequently have easily available to us – they come to mind more easily – when we take a more considered look. For example, if a page is loading and it is brightly coloured, that will prime a set of associations in our brains that are different to a page that is black and white or only has a very muted colour scheme.

Key concepts

The halo effect

The tendency for us to seek or be more likely to see positive qualities in something if we have had a positive first impression of it (or vice versa).

Visceral beauty

Our automatic gut feeling about the aesthetics of a visual design.

Thin slicing

The ability to draw general conclusions about something based only on a 'thin slice' of information about it.

Image complexity

The number of different details within a design that do not repeat.

Prototypicality

How well a design matches the typical format or structure of other designs in the same category.

Novelty can harm usability

If a site looks 'prototypical' – ie it looks like other sites, or looks like what we would expect it to – it is by definition non-novel. So there can be a tension between novelty and usability. An example of this is the uproar that always seems to follow any new design or user interface (UI) feature added to Facebook.

If a site is unusual looking, we may be unfamiliar with where to find information and how to interact with it – meaning that we might have to put more effort into using it. This is similar to the familiarity versus novelty issue described in Chapter 1.

The evidence suggests that a little bit of novelty can increase a site's attractiveness to users, but that too much probably leads to confusion. So an optimum balance would be ideal.

The neuro-packaging design expert

Supermarkets are like the commercial world's version of an art gallery. Tens of thousands of pack designs vie for our attention and approval as we walk along the aisles. Competition is intense. The front of a box of cereal may not have the cachet of a Cézanne. The design of a bag of nappies might not impress an art critic. Yet in their own ways they need to do something more difficult: make us stop, look at them and feel some kind of emotional attraction. All of this has to be done against intense competition, quickly and across millions of shoppers worldwide.

Given the importance of package visuals it is not surprising that companies such as Procter & Gamble (P&G) are keen to apply scientific insights into design. One example is their dedication to understanding first impressions, or, in their lingo, the 'first moment of truth'. It is that moment when we are standing in front of the supermarket shelves, our eyes flitting from one pack to another, deciding which to put in the trolley. It's in that moment that global brands are made.

P&G have learned how to optimize that first moment of truth: 19 of their brands have annual sales of over US$1 billion each. Whether it is the packaging of Gillette razors, Head & Shoulders shampoo or Ariel detergent, each is the product of thousands of man hours of design, research and testing to make them as visually arresting and tempting as possible.

▶

Dr Keith Ewart spent a number of years as a P&G senior insights manager working in corporate packaging. His job involved developing and hunting down new research techniques for better understanding package designs, particularly early on in the development cycle when new ideas were being born.

He began to realize that traditional ways of judging new pack ideas often stifled innovation and favoured functional over emotional solutions:

P&G designers and engineers are great prototypers and love to test and learn, but some of the more exciting ideas that teams had passion for would often be screened out because we would rate them based on rational criteria. Sometimes great ideas would be rejected against everyone's gut instinct. Knowing what I know now I think applying neuro design approaches like implicit-response testing could have been incredibly useful in assessing these early prototypes.

Good designers have the intuition and skills to design emotional experiences but, historically, insight managers have not had the tools to measure this emotional impact. The majority of quantitative market research is explicit or Type 2 based when, in reality, consumers see a product and have an automatic or Type 1 reaction. This is particularly important in-store or online when a consumer is browsing.

This decision is not a result of calculated, rational thought, yet we would test with rational, Type 2 testing. It was only when we started to understand behaviour – how people actually shop – and started testing more in context that we began to understand more. I believe that neuro design testing adds a vital layer to complement this understanding and is critical to understanding these first impressions.

Implications for designers

The main implication for web designers is that they need to remember that the first view of their home page will be disproportionately formative on users' opinions of the site, and whether they decide to linger and explore or click away. Whilst usability factors traditionally get a lot of attention in website development, it may feel unintuitive that the first impression could be just as important, if not more.

It is also important to consider the design conventions for the type of website you are creating. How do users expect the page to be constructed? Does your page look too different from those of its close competitors?

Not everyone will necessarily have the software or wherewithal to set up a proper first-impressions test that can flash up web images to participants at 50 or 100 milliseconds, but a quick-and-easy alternative is just to show each page or design for five seconds – this is a manageable time for opening and closing a page under controlled timing. There are even websites that will enable you to set up such tests automatically (eg http://fivesecondtest.com/ and https://www.usertesting.com/product/website-usability-testing). You can then ask users to give their ratings or opinions on what they thought of the design, how it made them feel, how trustworthy they thought the site is and so on.

However, one drawback to this approach is that you cannot be sure to what extent people are reacting to the general design of the page, versus the actual content. Most websites tend to keep their designs for a while, but change regularly the content of what appears within that design. Are people reacting to the design, or are they reacting to specific images or words that are currently in your design but may be changed in the near future?

An alternative solution,[24] which gets around this problem, is to use graphics software to put a low spatial filter on a screenshot of the webpage image. This effectively replicates what the human visual system is seeing within the first-impressions time window of around 0.05 seconds. It is equivalent to looking at a page whilst squinting. You see the general gist of the design but without being able to read the detail. Just showing people the page as a regular image means they could be biased by the particular content in how they report their reactions. Your content will presumably change more regularly than your overall design.

If there is the opportunity, it may also be worth considering the demographics of your users. In particular, their age, education level, gender and cultural background, as all these things can influence what creates a good first impression.

Clear at a glance

The types of images that perform well in first impressions are typically those that are clearly visible at low spatial frequencies. This is another way of saying that simple and clear imagery (as seen in Chapter 2) typically tends to have an advantage and be better liked (see Figure 4.5).

This is probably a good rule of thumb for any imagery that needs to be decoded quickly or recognized easily. For example, research has shown that icons (eg like those on smartphones) are more readily recognized if their designs are clear at low spatial frequencies rather than high ones.

Figure 4.5 Example of a webpage shown normally (left) and after being filtered by a low-pass (Gausian) filter set to 6.1 pixels (right).

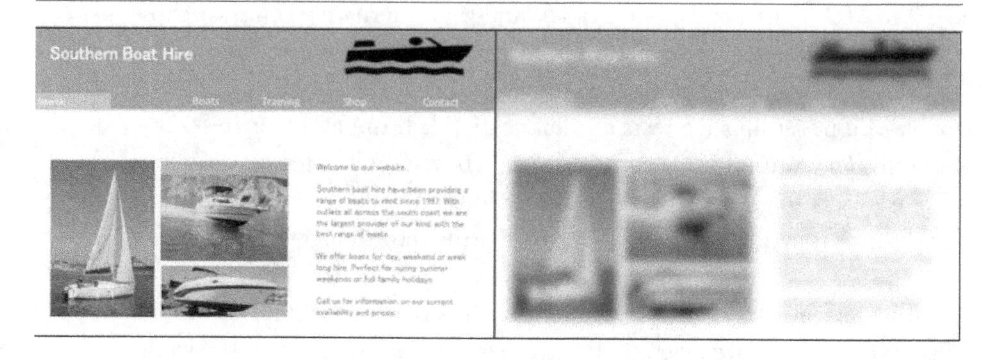

This also may have implications for a brand's distinctive visual assets and iconography, such as logos or key product imagery. Keeping images clear and simple helps their first impression but also helps them to be more readily identified.

The Mona Lisa effect

The enigmatic appeal of the *Mona Lisa* painting may be due to a neuro design effect called 'visuospatial resonance'.[25] It relies on the way that images can be rendered at different spatial levels: low spatial frequency images appear fuzzy when close up but can be seen from a distance, whereas high spatial frequency images are clear close up. The *Mona Lisa* has details in it at both spatial frequencies. When seen close up, she does not appear to be smiling. But there are details in the painting (some slight shading around the corners of the mouth) that are at a low spatial frequency that make her appear to slightly smile when seen at a distance or from the corner of your eye. It is called a gaze-dependent facial expression. This technique can be used to add intrigue to images that can be seen at different distances, such as posters.

Figure 4.6 Example of text at low and high spatial frequencies. When seen close up the word 'High' is clear, but when seen from a distance the word 'Low' is clearer.

SOURCE: Thanks to Dan Jones for this image.

CASE STUDY Website first impressions and the brain

In a study I worked on for software company Radware we showed 30 people three different retail websites loading in different ways. For example, one method was just loading as fast as possible.

Whilst they were viewing the pages we had fitted them with EEG sensors across their head to measure their moment-by-moment brain activity. One of the measures we were able to get was a reading of how emotionally attractive they found the pages.

The EEG brain data showed a statistically valid link between people's patterns of attraction to the page when it was still loading in the first five seconds, and their end reaction to the page after it had already fully loaded and been clearly visible on-screen for at least 10 seconds. Whilst this is only a relatively short-term effect, it shows that first impressions can form even before the page has fully loaded and people have even had a chance to consciously read and consider its content.

Summary

- Web users will form a rapid emotional like or dislike of a page within 0.05 seconds.
- The largest driver of first impressions appears to be the level of complexity of the page.
- Colours also play a significant role, although the role of both complexity and colour seems to vary according to age, culture, gender and education level of the user.
- When user testing webpage designs and layouts, it may be best to test for general aesthetic appeal by first putting the image through a low spatial filter to obscure the detail and replicate what people are seeing in the first 0.05 seconds.
- If unsure about the complexity of your page, and you don't have the opportunity to test it, consider erring on the side of slightly too simple rather than slightly too complex – the risks are less.

Notes

1. Tusche, A, Bode, S and Haynes, JD (2010) Neural responses to unattended products predict later consumer choices, *The Journal of Neuroscience*, **30** (23), pp 8024–31.
2. Nalini Ambady, N, Krabbenhoft, MA and Hogan, D (2006) The 30-sec sale: using thin-slice judgments, *Journal of Consumer Psychology*, **16** (1), pp 4–13.
3. Ambady, N and Rosenthal, R (1993) Half a minute: predicting teacher evaluations from thin slices of nonverbal behavior and physical attractiveness, *Journal of Personality and Social Psychology*, **64** (3), pp 431–41.
4. Banno, H and Saiki, J (2015) The processing speed of scene categorization at multiple levels of description: the superordinate advantage revisited, *Perception*, **44** (3), pp 269–88.
5. Lindgaard, G, Fernandes, G, Dudek, C and Brown, J (2006) Attention web designers: you have 50 milliseconds to make a good first impression!, *Behaviour & Information Technology*, **25** (2), pp 115–26.
6. Ekman, P (1992) An argument for basic emotions, *Cognition & Emotion*, **6** (3–4), pp 169–200.
7. Norman, DA (2004) Introduction to this special section on beauty, goodness, and usability, *Human–Computer Interaction*, **19** (4), pp 311–18.

8 Papachristos, E and Avouris, N (2011) Are first impressions about websites only related to visual appeal?, *Human–Computer Interaction (INTERACT 2011)*, pp 489–96.

9 Sillence, E, Briggs, P, Harris, P and Fishwick, L (2006) A framework for understanding trust factors in web-based health advice, *International Journal of Human–Computer Studies*, **64** (8), pp 697–713.

10 Hasan, Z, Gope, RC and Uddin, MN (2013) Do aesthetics matter in long-established trust?, *International Journal of Computer Applications*, **77** (13), pp 31–36.

11 Tuch, AN, Roth, SP, HornbaeK, K, Opwis, K and Bargas-Avila, JA (2012) Is beautiful really usable? Toward understanding the relation between usability, aesthetics, and affect in HCI, *Computers in Human Behavior*, **28** (5), pp 1596–607.

12 http://www.theguardian.com/lifeandstyle/2009/mar/07/first-impressions-snap-decisions-impulse (last accessed 25 August 2016).

13 Stecher, KB and Counts, S (2008) Thin slices of online profile attributes, in *Proceedings of the International Conference on Weblogs and Social Media*, AAAI Press (March).

14 Todorov, A and Porter, JM (2014) Misleading first impressions different for different facial images of the same person, *Psychological Science*, **25** (7), pp 1404–17.

15 Van der Land, SF, Willemsen, LM and Unkel, SA (2015) Are spectacles the female equivalent of beards for men? How wearing spectacles in a LinkedIn profile picture influences impressions of perceived credibility and job interview likelihood, *HCI in Business*, pp 175–84.

16 https://www.thinkwithgoogle.com/articles/creating-youtube-ads-that-break-through-in-a-skippable-world.html (last accessed 25 August 2016).

17 http://www.wired.com/2016/08/science-waiting-waiting-page-load/ (last accessed 25 August 2016).

18 Harrison, C, Yeo, Z and Hudson, SE (2010) Faster progress bars: manipulating perceived duration with visual augmentations, in *Proceedings of the SIGCHI Conference on Human Factors in Computing Systems*, pp 1545–8) ACM.

19 Reinecke, K, Yeh, T, Miratrix, L, Mardiko, R, Zhao, Y, Liu, J and Gajos KZ (2013) Predicting users' first impressions of website aesthetics with a quantification of perceived visual complexity and colorfulness, in *Proceedings of the SIGCHI Conference on Human Factors in Computing Systems*, pp 2049–58.

20 Reinecke, K and Gajos, KZ (2014) Quantifying visual preferences around the world, in *Proceedings of the 32nd Annual ACM Conference on Human Factors in Computing Systems* (ACM).

21 Tuch, AN, Presslaber, EE, Stöcklin, M, Opwis, K and Bargas-Avila, JA (2012) The role of visual complexity and prototypicality regarding first impression of websites: working towards understanding aesthetic judgments, *International Journal of Human–Computer Studies*, 70 (11), pp 794–811.

22 Quoted in: http://www.theabp.org.uk/news/the-psychology-of-good-website-design.aspx (last accessed 25 August 2016).

23 Tuch, AN, Presslaber, EE, Stöcklin, M, Opwis, K and Bargas-Avila, JA (2012) The role of visual complexity and prototypicality regarding first impression of websites: working towards understanding aesthetic judgments, *International Journal of Human–Computer Studies*, 70 (11), pp 794–811.

24 Thielsch, MT and Hirschfeld, G (2012) Spatial frequencies in aesthetic website evaluations: explaining how ultra-rapid evaluations are formed, *Ergonomics*, 55 (7), pp 731–42.

25 Setlur, V and Gooch, B (2004) Is that a smile?: gaze dependent facial expressions, in *Proceedings of the 3rd International Symposium on Non-Photorealistic Animation and Rendering*, pp 79–151) ACM (June).

Multisensory and emotional design

05

Figure 5.1 The Japanese have made cuteness into an art form called Kawaii. The images trigger our nurturing responses to babies with the visual cues of large eyes and heads

Louis Cheskin was a man ahead of his time. A Ukrainian-born psychologist, Cheskin worked in market research and made a number of significant contributions. For example, he helped turn margarine into a successful product by recommending its colour be changed from white to yellow. Based on his approach to design, Cheskin also correctly predicted (in contrast to the prevailing expectation of the time) that the Ford Edsel car would fail in the market, whilst the Ford Thunderbird would succeed (Figure 5.2). After this Henry Ford hired him.

Figure 5.2 A 1957 Ford Thunderbird

PHOTO BY USER: Morven at the Fabulous Fords Forever show at Knotts Berry Farm, Buena Park, California, USA on 17 April 2005; https://commons.wikimedia.org/wiki/File:1957_Ford_Thunderbird_white.jpg

Cheskin was ahead of his time because he understood that just asking consumers what they think about a design could be misleading. Instead of questionnaires, he preferred to rely on observational studies and experiments. For example, in one test he arranged for consumers to be sent three identical deodorant products each in a different colour. As he expected, consumers didn't realize they were identical and expressed preferences for one over the others.

The market research industry is only now embracing the importance of the non-conscious, and adopting research methods that measure non-conscious reactions. I suspect that if Cheskin were working today he would be using neuro-research methods.

Another of his insights was that people cannot help but transfer their feelings about a product's design or that of its packaging onto the product itself. He called this *sensation transference*. This is similar to the halo effect that we looked at in Chapter 4: initial positive impressions formed by a good design lead people to feel that a product or service is positive all round. It is also similar to findings from embodied cognition research, an area we will look at later in this chapter. For example, people will perceive an object as more significant if it is heavier. A smooth object can feel heavier because the smooth surface obliges us to grip it tighter, and we non-consciously associate a tighter grip with a heavier object. The sensory features of a design can drive our feelings about it.

As we saw in Chapter 1, digital technologies are spreading images like never before. Yet humans are multisensory creatures: touch, taste and smell are also important to us. Each is potentially a route to providing a pleasurable and emotionally engaging experience. Yet sadly, in our print-and-screen world, these channels of communication are not directly available to designers. However, there are indirect ways they can use them.

Activating other senses through visuals

Usually our brain blends together information from our different senses to give us a full idea of what we are seeing. For example, when we see someone speaking, we hear his or her voice at the same time as we see their mouth moving. The two streams of information – visual and sonic – match up; they are telling us the same thing. So our brain is surer of what is going on, because different senses are aligning, and the experience becomes stronger. Our brains are doing this all the time, it is called multisensory integration. The implication for design is that, if an image can suggest different sensory information that backs up what the visuals are depicting, it can convey a stronger experience to the brain.

Another aspect of multisensory integration is seen in the approximately 5 per cent of the population who have synaesthesia. This is where people's brains spontaneously mix together different sensory associations even if they are not really there. For example, seeing letters or numbers may trigger the experience of different colours or sounds in those with synaesthesia. Usually these associations are very personal to those who have synaesthesia. However, sometimes they can be more universal. For example, research shows that a number of qualities are linked in people's minds with colour. These include hot–cold, dry–wet, lively–numb and transparent–opaque.[1]

Do letters trigger colour associations?

An interesting example of synaesthesia is associating colours with letters. When those with letter–colour synaesthesia view black-and-white letters, areas in their visual brain that process colour become active.[2] This mixing up of letters and colours may occur because the areas in the brain that process letters and colours are close to each other. One study asked both children and adults to find particular letters, hidden inside one of two boxes whilst the boxes were covered over in different colours of fabric.[3] The experimenters recorded which box people reached into to find the letter.

They discovered the following associations:

- O associated with white;
- X associated with black;
- A associated with red;
- G associated with green.

These associations do not seem to be driven by the sounds of the letters. The experimenters know this because one group of children studied who were too young to read had these associations without knowing the sounds of the letters. Whereas word-meaning associations could have played a role in A being associated with red (A is for apples, and apples are often red) and G with green (G being the first letter of green). The two main drivers of colour–letter associations seem to be the letter shape and whether the colour word begins with that letter, or something highly associated with that letter is typically that colour.

Further research on colour–letter associations confirmed the 'A = red' association, not only for English speakers but for Dutch and Hindi speakers too.[4] The researchers recruited Dutch speakers in Amsterdam, native English speakers in California, and Indian-born native Hindi speakers who were living in California. They were asked to consciously assign colours to letters and days of the week. Whilst the participants often felt that their answers were random (suggesting they were not aware of these colour associations), patterns emerged in the results across people.

There was some agreement between the three language groups, with A for red and B for blue being the strongest pattern. Table 5.1 shows the most frequently chosen colours (along with the percentage of people who chose them) for each letter tested (sadly not every letter in the alphabet was tested).

Table 5.1 Letter–colour associations

Letter	Most frequent colour association
A	Red (60%)
B	Blue (30%)
D	Brown (24%)
E	Green (21%)
F	Red (26%)
H	Brown (13%)
K	Brown (13%)
I (capital I)	White (16%)
l (lower case L)	Yellow (21%)
N	Brown (15%)
S	Green (23%)
T	Green (22%)
U	Blue (18%)
W	White (15%)

One study found that more than 6 per cent of Americans with synaesthesia had colour–letter associations that matched a colourful set of letter fridge magnets made by toy company Fisher-Price that were popular in the 1970s and 1980s.[5] In other words, they may have non-consciously learned these connections between particular letters and colours during their childhood from repeatedly seeing them on their fridge (although, of course, it is also possible that the creator of the magnets also had a form of synaesthesia that matched those 6 per cent of Americans and hence had the same colour–letter associations!).

As well as letters being associated with colours, when using colour words, be aware of something called the 'Stroop' effect. This is simply the fact that it is easier to read colour words when the font is in the same colour as the word than when it is in a different colour.

Day–colour associations

The researchers in the previously mentioned study also tested the associations between days of the week and colours.

Sunday and Monday showed the greatest consistency, with Sunday being most associated with white or yellow across all three language groups, whilst Monday was either blue or red.

For English speakers, the most common colour associations with each day were (along with the percentage of people who had that association):

Monday = red (32 per cent)

Tuesday = yellow (27 per cent)

Wednesday = green (31 per cent)

Thursday = green (25 per cent)

Friday = blue (21 per cent)

Saturday = blue (22 per cent)

Sunday = white (20 per cent)

The fact that red is associated with both the letter A and Monday (which is often thought of and presented on calendars as the first day of the week) could be a linking between the letter/day that comes first, and the colour that first comes to people's minds. It may be that red is just the most easy-to-think-of colour.

Do shapes have colours?

Other research has looked at associations between shapes and colours.[6] The results showed the strongest associations being of triangles with yellow, and between circles and squares with red.

Analysing the results revealed that the link of shapes to colour was being influenced by two types of colour associations: the warmth or coolness of the colours, and their brightness. For example, red is associated with warmth, whilst green and blue are associated with coolness. Yellow is naturally bright, whereas black naturally dark. Circles and triangles were perceived as warm, the rhombus as cool. Squares as dark.

Our preferences for colours and shapes seem to be intertwined and connected to the sensory and emotional associations we have with them. These associations are not absolute, but rather trends that we see across populations. There is still individual variation between people.

Shapes can also boost our associations with different sensory information. Colour can affect our perception of taste. For example, red tends to enhance our perception of sweetness, whilst green weakens it. This is because we evolved to recognize the ripeness of fruit. Curves and circles can also trigger our associations with sweetness, whereas spikey shapes trigger the feeling of bitterness. An example of the latter is the red star on bottles of San Pellegrino sparkling water (sparkling water being somewhat bitter). It appears that our senses are deeply interconnected in the brain. Whilst people with synaesthesia are consciously aware of this, we may all be making these connections at a non-conscious level.

Food photography is another example of how images can evoke different senses. Food photography in ads and packaging has become more sophisticated in recent years, with greater care being taken to compose the photographs and more use of high-definition images. Effective food photography uses techniques such as:

- Use of high definition: high-definition images can reveal appealing details of food, showing greater texture, crispiness, sugar coatings, more subtle colours, fizzy bubbles in drinks, and steam rising from hot food or drinks.
- Signs of freshness: the texture of the food should look fresh, avoiding anything that would suggest it has been standing around or – for hot food – has gone cold (steam can help show that). With fruit or vegetables a moist, glistening appearance can help suggest freshness.
- Suggesting the food is ready to eat: things that suggest the food is ready to eat help drive appetite appeal. Movement is one way: showing a drink being poured, or cream being poured onto a dessert, for example. Also showing a knife, fork or spoon either next to the food, or digging into the food.

Colours also can become associated with different brands or product categories, through repetitively seeing them. For example, light blues are often used for detergent or hand-cleaning products. Using a tone of blue on a food product could, in this example, trigger non-conscious associations with detergents that harm appetite appeal. Taking the particular colour you are considering using and using it in a Google image search can perform a quick risk check. For example, when choosing the colour of packaging for a food product, you would probably want to avoid a hue that was frequently used in detergent or cleaning products. Such connections are learned in consumers' non-conscious minds and could disrupt appetite appeal.

Colour

Colours do not really exist objectively; they come about as a result of the way our brains interpret light. As Sir Isaac Newton put it: 'For the Rays, to speak properly, have no Colour. In them there is nothing else than a certain power and disposition to stir up a sensation of this Colour or that.'[7]

Whether we experience colours differently is a question that has been pondered by philosophers for centuries. You have probably had the experience of pointing to something and calling it one colour whilst a companion insists it is a different colour. There was a famous philosophy paper published in the 1970s called 'What is it like to be a bat?' that points out that, whilst we all inhabit the same world, we can never know for sure how others experience sensations, such as colour.[8] We can measure others' brains from the outside, but never truly get inside them.

Culture also teaches us to be sensitive to different colours. For example, some scholars have found a lack of the word 'blue' in ancient texts such as *The Odyssey*, leading them to speculate whether seeing blue might not be universal at all. Intriguingly, an experiment with the Himba tribe in Namibia – who do not have a word for 'blue' – showed that they had difficulty in spotting a blue square amongst a selection of green squares. Yet they were adept at spotting a slightly different shade of green square amongst a series of green squares, something that most Westerners found very hard to do.[9]

The chromostereopic effect

Different colours – ie wavelengths of light – hit our eyes in different ways. Due to this effect it is hard for us to focus on both blue and red when they are next to each other. This makes this particular colour combination feel uncomfortable to look at. It also makes the colour 'cobalt blue' hard to look at (because the colour is a mixture of red and blue). Another aspect of this is that red objects or text on a blue background will appear to float over the top of it (although the exact opposite perception is experienced by some people). The floating effect can also occur with yellow on blue, and green on red. In other words, the general rule is that if you mix these colours to make something stand out (float above) it is usually best to place warmer colours on top of cooler colours.

Colours obviously affect us emotionally. Young children often place importance on deciding on their favourite colour; when making expensive purchases like buying a new car, its colour is often a critical variable; and people decorate the walls of their homes in colours chosen to create a particular feel and mood. Most people understand, for example, that a cold room – such as one that gets little direct sunlight – can be warmed up by painting it in a warm colour such as orange or red. Preferences for particular colours seem highly personal, but are there any universal patterns?

Research shows that Westerners generally prefer cool colours (such as green and blue) over warm colours (such as yellow and red).[10] Overall, blue is the colour most universally liked across cultures, perhaps because of its association with skies and bodies of water. Equally, dark yellow is disliked across different cultures. Conversely, longer-wavelength colours such as red evoke higher levels of emotional excitement. So there are two separate effects that colours have: an evaluative or liking effect, and an excitation effect.

Red in particular has a universal effect that likely has an evolutionary explanation. Red is a sign of extreme heat, and of bleeding. Red is regularly used as a warning colour, for example on signs telling people to stop, or warning them of danger. There is also evidence that people act weaker when facing an opponent wearing the colour red. Research shows that Olympic competitors in combat sports are more likely to win if they are wearing red than blue.[11]

Submission in the presence of the colour red is, intriguingly, also found in monkeys. In one study, a male and female researcher entered a colony of rhesus macaque monkeys, and placed a slice of apple in front of them in the presence of the monkeys. Most of the time the monkey would move forward and take the apple. On different occasions the researchers would be wearing either red, blue or green clothes. When they were wearing red, the monkeys would not take the apple from in front of them. It made no difference if the researcher was male or female, the same submission to red was seen.[12] Submission to red may be an evolutionary by-product of recognizing when a person's face is flushed red with anger, hence being more likely to attack.

This effect may also appear in situations in which we are being judged, even if we are not directly facing-off against another person. One study found that exposure to the colour red impaired people's performance in an IQ test.[13] They hypothesized that red reminds us of the possibility of failure, and hence makes us more nervous about competing or trying our hardest. Designer David Kadavy, in his book *Design for Hackers* (2011), speculates whether this effect could also make us less rational and more open to emotive appeals in a retail context.[14] He observes how the US retailer Target – that uses red extensively throughout its shops – seems to make people susceptible to spending more than they intended to. Perhaps red also makes us more 'submissive' to a retailer's claims (eg red signs highlighting a special offer) or dampens our rational thinking and encourages a more emotional response.

Another type of colour vision

For a long time scientists have known of the rod and cone cells in our eyes that sense light and allow us to see. Rods are very light sensitive and we rely on them entirely to see in low light conditions. However, they cannot perceive colour. This is why at night things look more black and white. There are three types of cone cells: one most sensitive to blue light, one to green and one to red. Comparing the inputs from these allows our brains to compute the colours of what we are seeing. We have about 120 million rod cells and 6 million cone cells in each eye.

However, a new type of light-sensitive cell was discovered in the late 1990s. The melanopsin-expressing cells are sensitive to blue light, but rather than help form our conscious visual perception of what we are seeing, they feed into our brain's body clock and areas involved in emotions. The sunrise in the morning, or sunnier weather in the summer, cause hormonal changes that make us feel more positive and energized than in the middle of the night or the lower light of winter.

Scientists used to think that it was just the amount or intensity of light alone that regulated our sleep-and-waking rhythm. However, now it is looking like it could be whether we are exposed more to blue light, or its opposite: orangey-yellow. Whilst yellow light dominated our ancestors' vision around sunrise or later in the afternoon (when they would have been more active) as the sun was lower, it was bluer at night and midday when they either slept or laid low to avoid the most harsh UV light. This could explain why we associate yellow and orange colours with feeling positively energized (for example, the classic smiley-face icon is typically yellow) and blue with feeling more subdued and mellow.[15]

Colour blindness

Around 8 per cent of men and 0.5 per cent of women are colour blind. There are different forms of colour blindness, but the most common type is a reduced sensitivity to distinguishing between colours that are made up of red and green. Effectively this means that many colours will be hard to distinguish for colour-blind people.

When creating designs, particularly those that need to be effective with men, it may be worth checking your colour combinations. There are colour-blindness simulators online (such as: http://www.color-blindness.com/coblis-color-blindness-simulator/) where you can upload images and see how they will look to people with different forms of colour blindness.

Embodied cognition

Sometimes consumers complain that a product feels 'too light' (the iPhone 5 had this criticism levelled at it). This is a slightly strange objection: why should consumers find a lightweight product objectionable? Isn't lightness a benefit, making something easier to carry? The clue is in the very word lightweight. We use lightweight as a metaphor for someone or something that is lacking in worth and importance. Our brains often use such metaphors – originating in the understanding we gain of the world from our senses and bodily movements. Psychologists call this form of intuitive thinking *embodied cognition*.

If we think about it, drawing conclusions about brands and products on the basis of such sensory metaphors is a bit irrational. After all, discounting a product just because it feels too light typically serves no purpose. However, such connections are a shortcut, non-conscious form of reaction that most of us have a lot of the time.

Different physical properties like cold, warmth, roughness, smoothness, lightness and heaviness get mapped non-consciously on to a range of associations. Apple products and stores use materials that feel smooth, presumably to convey the smooth and easy experience of using their products.

Even gestures when interacting with a website could produce embodied cognition associations. For example, on smartphones or tablets, including gestures of swiping an object towards the person (ie down on the screen) could make them like it more, as moving things towards us is a gesture of acceptance.

Word sound associations

Many words have roots that can be traced back at least 8,000 years, and probably came about due to the way that they sound and the shapes our mouths make when we say them.[16] For example, the word 'open' has in it the sound '*pe*'. When we purse our lips together and then open them, our mouth is performing a movement in making the sound that mirrors the meaning of the sound. Similarly saying '*mei*' moves your lips into a smile, and the sound is at the root of the word 'smile'. This is probably how many of our words originally formed, and it makes words feel intuitively right. In a sense, these prehistoric syllables are a product of embodied cognition as they map the movement of our bodies on to the meanings we are trying to convey.

Table 5.2 shows some examples of these ancient word sounds that give our modern words an intuitive feel.

Table 5.2 How words evolved from sound meanings

Sound	Meaning	Example Modern Words
Ak	Fast or sharp	Acrobat, acute, equine, acid
An	To breathe (and to be alive)	Man, animated, animal
Em	To buy	Emporium, premium
Kard	Heart	Cardiac, Courage
Luh	To shine	Lunar, lustre, luxury
Mal	Dirty	Malady, Malign, malaria, melancholy
Mei	To smile	Miracle, marvellous, smirk
Min	Small	Minimum, mite, minus
Prei	First	Prize, praise, prime
Re	To go backwards	Reverse, retro, rear, rescue
Spek	To see	Spectacles, inspector
War	To guard	Wardrobe, warden, beware

Awareness that certain sounds can create an instinctive association can be useful for picking key words to create emotional or meaning associations. Designers and marketers picking words that will be important, such as creating new brand or product names, or use of words in headlines, should: 1) think about what emotional or meaning associations the word conjures up; 2) pay attention to the way that your mouth and face move when you say the word. Is your mouth moving into a smile? Are your lips pursed together or spread widely apart?

The *affect* heuristic

Heuristics are mental shortcuts that we use in making decisions. In Chapter 7 we will look at more examples of heuristics, but there is one that is directly relevant to emotional engagement. The *affect* heuristic refers to situations in which people make decisions using the mental shortcut of how they feel, even if, strictly speaking, this may be irrational. In everyday language we say that someone is going with their gut feeling. This is related to the first impression and halo effects that we looked at in Chapter 4. Just creating a positive feeling can be enough to bias someone's choice.

Emotional stimuli can bias our decisions without us being consciously aware of them. In one experiment, people were sat in front of a screen and a smiling or frowning face or a neutral shape was flashed up for just 1/250th of a second – enough for the image to be perceived at a non-conscious level only. Participants were subsequently shown Chinese ideograph characters and asked to rate how much they liked them. When shown after non-consciously seeing a smiling face they were more likely to prefer the image than if shown after a frowning face or a neutral shape.[17]

The affect heuristic highlights the importance of creating a positive gut feeling about a design, using elements – like smiling faces – that are likely to make people feel good. Also showing people interacting with your product or service can help drive emotional engagement. We are most sensitive to things touching our hands or face, so images of things in contact with someone's hands or face may be more sensorially evocative and emotionally engaging.

Faces

One of the easiest and most effective ways of creating an emotional response through imagery is with faces. Our brains have dedicated regions for processing faces, and we can detect emotion on a face within 100 milliseconds. Our visual brains are so sensitive to spotting faces that we sometimes see them when they are not there, a phenomenon known as *facial pareidolia* – for example, seeing faces in cloud or rock formations. This could be an unavoidable consequence of any system that is sensitive to detecting faces as it is also seen in software that is designed to detect faces.

Faces will work better on larger formats than smaller. On smartphone screens facial expressions are less clear. This is one of the reasons why, despite people spending more time looking at mobile screens, TV and cinema screens are still ideal for ads that aim to emotionally engage us. In studies I have performed using electroencephalography (EEG) on TV ads, I have seen a trend for viewers being emotionally engaged towards faces that are oriented towards the camera, then becoming disengaged when they turn away.

In 1973 American statistician Herman Chernoff suggested leveraging our sensitivity to faces in order to illustrate data. When trying to depict variations from an average value, Chernoff suggested taking a standard line drawing of a face and varying features such as the shape of the face, the mouth, the distance between the eyes and so on. For example, rather than use a graph, you could show the difference in the numbers you are illustrating by varying the width or length of a series of otherwise identical cartoon faces.

Emojis

Emojis are a throwback to the pictographic origins of writing. It was not until the ancient Greeks that abstract letters were organized into a written language system. Before that, picture systems such as Egyptian hieroglyphics were used. Emoji characters were first created in the 1990s by a Japanese mobile-phone provider. The word itself is a combination of the Japanese words for picture (e) and character (moji). As we are so good at recognizing emotional expressions in faces, emojis are an excellent shortcut to emotional expression, without the need to hunt for the right word to express the tone or context to our message. Vyvyan Evans, Professor of Linguistics at Bangor University, has suggested that emoji characters are now the world's fastest-growing language.[18]

The uncanny valley

People like looking at faces, but many feel a visceral discomfort at seeing artificial faces, such as certain computer-generated faces or robot faces. The uncanny valley effect is often particularly driven by artificial eyes, which can look oddly cold and dead. We effortlessly process a lot of information when watching people's faces, and if everything isn't perfect, we can spot it easily. For example, lack of movement in the upper half of a face when a person is talking can make the face look unnatural.

A similar effect may occur with computer-generated imagery in general. People often complain that today's movies with special effects paradoxically look worse than those from similar movies made in the 1990s. If computer graphics and processing speed have improved, today's effects should look better, yet they often create a feeling of less realism and impact. This may be because when digital computer effects were in their infancy, film-makers were limited in how much they could use them: for example, compositing a digital computer-generated dinosaur into a real filmed background in the movie *Jurassic Park*, whereas now they also have the capability to make completely computer-generated backgrounds as well as characters. Hence, in some shots, nothing is actually 'real'. Whilst people may have difficultly consciously articulating why a shot looks real or not, they may still be left non-consciously feeling that something is fake.

The neuro movie-analyst

Ernest Garrett is passionate about using evolutionary psychology to understand what makes a movie a hit or a flop. As a successful script reader in Hollywood for a number of years it was his job to read thousands of prospective screenplays and novels to find the ones that showed the most potential for becoming popular movies.

He now runs a popular YouTube channel – StoryBrain (https://www.youtube.com/user/StoryBrain) – explaining his theories, based on the inherent things that appeal to our brains:

> *Our brains are extremely sensitive to detecting visual images that just don't look real, which has become a problem in Hollywood as more movies are relying on filling the frame with entirely computer-generated imagery. It has led to the weird trend for audiences to find many recent movie-effects less believable than those that are 20 or 30 years old.*

▶

Back then film-makers were constrained more in what images they could create on computers, so they tended to blend CGI characters into real backgrounds. Nowadays they can even create artificial backgrounds and the audiences' eyes pick up the lack of real-world cues. Even if the artists have done a good job the result can sometimes look too perfect, and contrived. One good example of an old movie that did special effects well was Honey, I Shrunk the Kids. *It was about a group of children who were shrunk down to the size of ants. Our brains haven't evolved to sense how things move at that scale – in the real world, because of air resistance, ants can drop from a great height and not get hurt. But our brains have evolved to sense gravity and weight at a larger size. That movie had effects that felt natural to us as it showed a lot of gravity cues from our everyday world but set in miniature.*

When one of his videos – explaining the way that overuse of CGI can feel wrong to our brains – went viral on YouTube (getting over 1 million views), there were rumours that the director of *Star Wars VII* saw it. Just after the video went viral JJ Abrams went out of his way at a press conference to emphasize that his new Star Wars film would not overrely on CGI.

Cuteness

Animals and humans have evolved to find the sight of babies cute and adorable to ensure that we protect and nurture them. Baby animals that are born relatively helpless and needing a long period of nurturing look cute to us (such as dogs, rabbits or bears), whereas those born ready to go (such as fish and insects) don't.

Chubby limbs and large eyes are two of the triggers to cuteness. Babies' heads are disproportionately large (compared to adults), so large heads, in particular large foreheads, chubby cheeks, soft skin and large eyes help drive cuteness. Designers and artists have long co-opted these cuteness drivers into the design of things like mascots and even products. Car fronts are often redolent of faces, and cars like the Mini use large round headlights to mimic the large eyes of babies, making them look cute and friendly. Even features like curves and smooth surfaces may help evoke similar feelings of friendliness and cuteness.

Kawaii: the Japanese cuteness aesthetic

The Japanese have made cuteness into an art form. They call it Kawaii, which refers to cuteness, but originally meant a flushed (ie blushing) face. Kawaii is seen in the popularity of cute cartoons that look like small animals, such as the Pokemon or Hello Kitty characters, with faces that are exaggerated to accentuate their childlike look, such as having oversized eyes. Kawaii is essentially the peak shift principle applied to the cuteness of young faces.

This aesthetic – of making things look cute – is aimed at adults as well as children.

Anthropomorphic design

The human visual system is, as we have seen, very good at detecting patterns. In fact, it is so sensitive to patterns that it will often think it sees a meaningful pattern when there is none.

We are particularly attuned to seeing people and faces. A good example of this is the Johansson effect that we looked at in Chapter 2: our ability to clearly see the movement and shape of a human body even when all we can see are a series of white dots attached to its limbs.

Anthropomorphic imagery can add personality and emotional engagement to a design – adding friendliness or just greater emotional engagement. Anthropomorphic effects are used by brands when they create a mascot or character that communicates the brand personality, such as Ronald McDonald or the Energizer bunny.

Curved and pointy shapes

People generally have an instinctive preference for curves over sharp or angular designs. Sharp shapes signify objects that could hurt us, so it's natural that our visual brains instinctively avoid them. For example, one study showed participants 140 pairs of objects, such as watches or sofas, letters and abstract shapes. Each pair was the same type of object but one was more angular, one curvier. They were also shown equivalent images that were neutral – neither curvy nor angular. People were significantly more likely to prefer the curvy designs over the neutral ones, and significantly less likely to prefer the angular ones.[19]

In one study, babies and adults were sat in front of computer screens and shown lots of pairs of simple image shapes whilst an eye-tracker monitored which one they looked at first, and how long they spent looking at each one. Both adults and babies were more likely to look at shapes with tapered or curved lines than those with straight ones, and the adults were more likely to spend longer looking at them. A further study was conducted where people looked at shapes whilst their brains were being scanned in an fMRI scanner. They found a distinct pattern of activity when people looked at the curved/tapered shapes compared to the straight-lined ones that was comparable to a similar study that looked at brain activity in macaque monkeys. The similarity of activity in humans and monkeys, and the fact that babies, as well as adults, looked at the curved/tapered shapes first, suggests that this is a fairly primitive, universal visual brain preference and not something that people have learned from their particular cultural upbringing.[20] In design, curves can feel approachable, sharp angles and spikes can feel unfriendly. For example, the McDonald's arches convey a feeling of comfort and nurturing.

The peak shift effect that we covered in Chapter 2 can be useful in evoking sensory qualities in an image. The idea is that the designer should think about the unique visual features that create the sensory or emotional effect they want to convey – and enhance them. For example, if the aim is to enhance the friendliness of a design, look for ways to add in exaggerated curves.

Visual saliency may be at least partially driving our tendency to look at curved and tapered shapes before straight-lined ones. Curved or tapered shapes are more unusual than regular ones, so they seem more unlikely, and more likely to tell us something interesting about what we are seeing – so our eyes are drawn to them.

Key concepts

Multisensory integration

The fact that our brains blend together different sources of sensory information to create a full understanding of what we are seeing.

Synaesthesia

A phenomenon of some people naturally having a conscious experience of one sense being activated by another, such as them experiencing colours when they see shapes.

Embodied cognition

The way that we use feedback from our bodies to influence our thinking, such as using weight as a metaphor for things that are more significant.

The affect heuristic

The mental shortcut of people making decisions based on how they feel rather than their rational appraisals.

However, in the real world there are other factors in a design other than just curviness. For example, whether a design is symmetrical or well balanced – as we have seen in previous chapters. Or how much expertise and visual sophistication the viewer has (eg whether they are naturally interested in art, have taken art classes and so on). One experiment that compared the effects of these factors found that people with less expertise were more likely to prefer curved shapes, whilst those with more expertise had no significant preference between the curved and angular shapes. However, a replication of the experiment revealed the opposite pattern! It may be that the expertise of the viewer interacts with their preference for angularity in ways we don't yet understand. Yet given that most people are not experts, and most designs have to 'work' with non-experts, it is probably worth betting on the weight of evidence across studies for a preference for curves.[21]

Things that look smooth and curvaceous draw us in with an instinctive urge to touch and hold them. An example is the design of the classic curvy Coca-Cola bottle. As designer Raymond Loewy described it: 'Even when wet and cold, its twin-sphered body offers a delightful valley for the friendly fold of one's hand, a feel that is cosy and luscious.'[22]

As well as spiky shapes, there are a number of images that can create an instinctive, reflexive emotional withdrawal effect. Anything that represents a threat – be it a scary animal, smashing glass, etc – could lessen the emotional attractiveness of a design. This might sound silly, as we don't like to think of ourselves being put off by images like that, and we might not even be consciously aware of it.

Summary

- Multisensory integration is the term to describe when more than one channel of sensory information – such as sight and sound – is feeding our brains the same information.
- Letters, shapes, numbers and even days of the week can evoke colour associations.
- There is some evidence for a general preference for 'cooler' colours (eg greens, blues) over warmer ones (reds, yellows).
- Generally, people prefer curved shapes over angular or spiky ones.
- Faces are an effective way to drive emotional engagement in a design.

Notes

1 Albertazzi, L, Koenderink, JJ and van Doorn, A (2015) Chromatic dimensions earthy, watery, airy, and fiery, *Perception*, **44** (10), pp 1153–78.

2 Hubbard, EM, Arman, AC, Ramachandran, VS and Boynton, GM (2005) Individual differences among grapheme-color synesthetes: brain-behavior correlations, *Neuron*, **45** (6), pp 975–85.

3 Spector, F and Maurer, D (2008) The colour of Os: naturally biased associations between shape and colour, *Perception*, **37** (6), pp 841–47.

4 Rouw, R, Case, L, Gosavi, R and Ramachandran, V (2014) Color associations for days and letters across different languages, *Frontiers in Psychology*, **5**, pp 1–17.

5 http://blogs.discovermagazine.com/d-brief/2015/03/04/synesthesia-based-alphabet-magnets/#.V4JBg_R4WnO (last accessed 25August 2016).

6 Albertazzi, L, Da Pos, O, Canal, L, Micciolo, R, Malfatti, M and Vescovi, M (2013) The hue of shapes, *Journal of Experimental Psychology: Human perception and performance*, **39** (1), p 37.

7 Quoted in Zeki, S (1999) Splendours and miseries of the brain, *Philosophical Transactions of the Royal Society of London B, Biological Sciences*, **354** (1392), pp 2053–65.

8 Nagel, T (1974) What is it like to be a bat?, *The Philosophical Review*, **83** (4), pp 435–50.

9 http://www.iflscience.com/brain/when-did-humans-start-see-color-blue (last accessed 25August 2016).

10 Palmer, SE, Schloss, KB and Sammartino, J (2013) Visual aesthetics and human preference, *Annual Review of Psychology*, **64**, pp 77–107.

11 Hill, RA and Barton, RA (2005) Psychology: red enhances human performance in contests, *Nature*, **435** (7040), p 293.

12 Khan, SA, Levine, WJ, Dobson, SD and Kralik, JD (2011) Red signals dominance in male rhesus macaques, *Psychological Science*, **22** (8), pp 1001–03.

13 Elliot, AJ, Maier, MA, Moller, AC, Friedman, R and Meinhardt, J (2007) Color and psychological functioning: the effect of red on performance attainment, *Journal of Experimental Psychology: General*, **136** (1), p 154.

14 Kadavy, D (2011) *Design for Hackers: Reverse engineering beauty*, John Wiley & Sons, London.

15 http://www.livescience.com/21275-color-red-blue-scientists.html (last accessed 25August 2016).

16 Stevens, C (2014) *Written in Stone*, Penguin Random House, London. Interestingly, if a speaker of one of the Indo-European languages was able to travel back in time to the Stone Age era, many words would be recognizable to them, such as thaw, path, halp (help), lig (lick), mur (murmur), ieh (yes), neh (no).

17 Winkielman, P, Zajonc, RB and Schwarz, N (1997) Subliminal affective priming resists attributional interventions, *Cognition & Emotion*, **11** (4), pp 433–65.

18 Lucas, G (2016) *The Story of Emoji*, Prestel Verlag, London.

19 Bar, M and Neta, M (2006) Humans prefer curved visual objects, *Psychological Science*, **17** (8), pp 645–48. See also: Bar, M and Neta, M (2007) Visual elements of subjective preference modulate amygdala activation, *Neuropsychologia*, **45** (10), pp 2191–200.

20 Amir, O, Biederman, I and Hayworth, KJ (2011) The neural basis for shape preferences, *Vision Research*, **51** (20), pp 2198–206.

21 Silvia, PJ and Barona, CM (2009) Do people prefer curved objects? Angularity, expertise, and aesthetic preference, *Empirical Studies of the Arts*, **27** (1), pp 25–42.

22 Loewy, R (1951) *Never Leave Well Enough Alone*, Simon and Schuster, New York, p 297.

Visual saliency maps

06

Figure 6.1 Viewers' eyes are immediately drawn to things that are different to their surroundings, a property that neuroscientists call 'visual saliency'

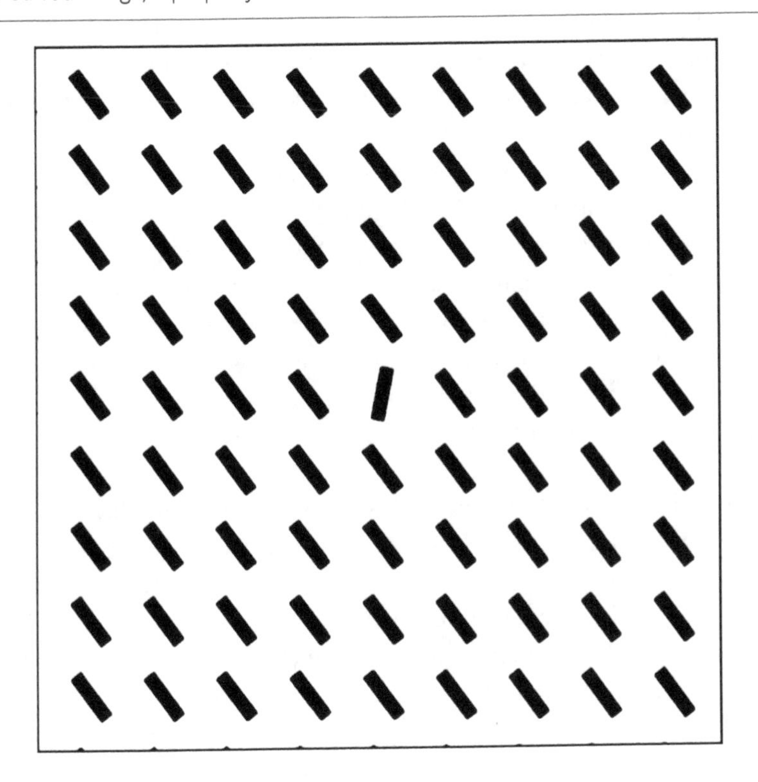

In 1967 a secret club, not open to the general public, was opened in Disneyland. The club was unusual as it was the only place in the Magic Kingdom to serve alcohol. Walt Disney himself initially resisted the idea of the club, but was under pressure from General Electric (GE), one of the original corporate sponsors of Disneyland. Walt wanted to bring into his park some of the attractions his Imagineers had created for the 1964 New York World Fair. As a sponsor, GE wanted a VIP corporate lounge, of the sort that the World Fair had hosted. Eventually Walt relented and 'Club 33' was built.

Located in the New Orleans Square, Club 33 is hidden behind an innocuous-looking door and is missed by most visitors who pass by without noticing it, almost as though a spell of magical concealment exists around it. It is not open to the average visitor, only for guests who must join a five-year waiting list, and pay tens of thousands of dollars for the privilege of frequenting it.

The truth to its near invisibility is colour. Disney designers have created two colours that help to make the more unsightly elements of the park visually disappear. They call them 'go away green' and 'no seeum grey'![1] The Magic Kingdom needs to maintain its aura of beauty and enchantment, but as with any major theme park it needs to contain many buildings and structures that look more industrial or mundane. For example, administration and utility buildings, or the rear sides of buildings that are not covered with a fairytale facade. With a coat of the right colour these lacklustre structures simply fall out of sight.

Club 33 isn't ugly, in fact its interior is decorated with a Southern old-world opulence. However, the park doesn't want to draw attention to it.

Designers and artists have often been requested to hide things: both Picasso and Matisse were commissioned by their country's naval authority to devise camouflage patterns. However, the challenge faced by the Disney designers, to conceal certain areas of the park, is in a sense the direct opposite of that faced by most design projects: how to grab attention and be noticed. Whether it is packaging in a visually competitive shelf display, a print ad in a magazine, or an important button on a webpage, understanding why some design features grab more attention than others is important.

How we decide where to look

Our visual system is now one of the best-understood processes in the brain, and neuroscientists have made great progress in modelling what captures people's attention. Computer simulations of our visual system are probably now the most advanced and realistic of any of our brain functions.[2] As visual information from our eyes enters our brains, two types of processing occur more or less in parallel.

First, there is bottom-up processing of the raw visual elements: the colours, contours, contrasts, luminosity, movements and textures of what is in front of us. Before our brains even recognize what we are seeing, the basic visual features have to be decoded. Then, almost immediately, our brains engage in top-down processing. Based on our memory, we begin to categorize what we think we are seeing. It might be a face, a car, a human figure. The sooner this top-down information can be fed into the mix, the sooner we can understand what we are seeing, which also helps in the processing of the raw visual elements. For example, we don't always see things in the clearest, most ideal viewing conditions. If we are looking at something in dim light, or something that is partially obscured, it may not be immediately obvious what we are seeing. We can see a series of lines, colours and shapes, but how do they fit together? Our brain needs to figure this out quickly or we would often be left completely confused by our eyes. If our top-down processes can suggest what it is likely to be, putting together the raw bits of information – shape outlines, textures, etc – becomes easier, like piecing together a jigsaw puzzle with the help of the overall picture. In other words, the top-down experience and memory processes are in constant communication with the bottom-up nuts-and-bolts processing of raw information.

Bottom-up processes are largely automatic and inbuilt in our brains. Neuroscientists call this stage of visual processing *pre-attentive* because it occurs before we are properly paying focused attention and aware of what we are looking at.

Top-down processes are more memory, context and goal related. For an example of this, see Figure 6.2.

Figure 6.2 Top-down versus bottom-up visual processing

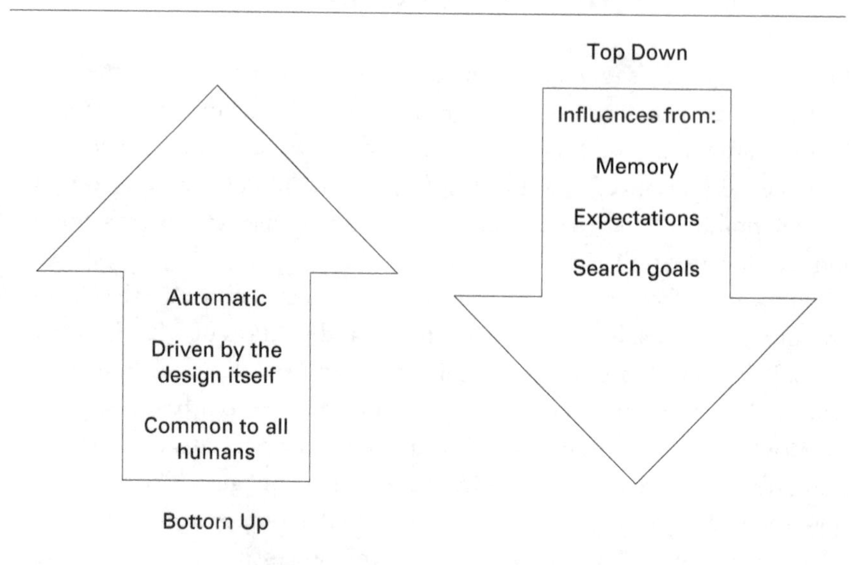

However, as our brains process the visual field in front of us they can only do so in detail one area at a time. We can only see detail in a relatively small area at any one time, as our eyes have a small field of detailed high-definition vision, surrounded by a larger field of low-definition vision. Our eyes make a series of quick movements, or *saccades*, in order to figure out what we are seeing and fill in the details.

Yet if we can only see details of things one at a time, there has to be a very quick and automatic way to *prioritize* what to look at first. For example, if we didn't prioritize places for our eyes to look, and our gaze pattern was completely neutral, we might always start by looking at the top left of what is in front of us, scan our eyes along, move down, scan along etc. But we don't look at things in this way. Instead, our brains have evolved to prioritize for things that move, or things that stand out against their surroundings. If something is moving, it could be a threat approaching us. If something is different from whatever is surrounding it, it is interesting because it is different. Again, it could be a threat, or it could be a bright-coloured berry standing out as different from the background foliage. These pressures helped to sculpt the brains of our hunter-gatherer ancestors.

Neuroscientists call this quality of something being different from its background 'visual salience'. In the mid-1980s, two neuroscientists at the California Institute of Technology introduced the idea that our brains produce a visual saliency map. Our brains are continuously mapping the colours, contours, shades, luminosity, etc of our surroundings, and then these elements are blended together and our visual system looks for areas that appear different to what is around them. Maybe they are moving, or have a contrasting colour, or appear brighter. The visual system then tells our eyes to move to focus on that area to see it in closer detail. Items that are visually salient get looked at earlier, more often, and for longer.[3]

Evolutionary pressures mean that it is important for animals to monitor their environments constantly for threats and opportunities. As they are driven by the mechanics of how our visual cortex processes incoming information, these mechanisms are common to all humans. Whilst movement is particularly attention-grabbing, for still images the three most important features for saliency are colour (including brightness), patterns (things like the orientations of lines) and size. Other visual features that can also affect saliency include depth (or cues in a design that imply depth), shape, flicker, and something called the *vernier offset* (the degree to which two or more different lines or contours line up with one another).[4]

This process needs to happen quickly, in order that we can detect threats (eg predators) or catch opportunities (eg prey, or unusual things we can learn from). It also needs to happen automatically. Think of how exhausting it would become if during every waking second we had to deliberately decide what to focus our eyes on first. Therefore this aspect of visual saliency is bottom-up and System 1.

However, top-down processing can also affect our saliency map: if we recognize an object as something we are interested in - for example a face, or food. Its lower-level visual features may not be that different from their surroundings, but they become salient to us because we are inherently interested in them. Also, if we are searching for something, such as looking for a can of Coca-Cola in a supermarket, our top-down processes will send signals to the bottom-up system to become sensitive to things like the colour red. Lastly, our top-down processes have expectations for the types of things we expect to see in whatever our environment is. If an object looks out of place simply because we don't expect to see it, it can become salient.

Equally, our top-down processing can suppress things even if they are salient from a bottom-up perspective. For example, if we are concentrating on searching for something that we know has a particular colour or shape, such as our favourite brand of product in a supermarket, then we visually supress things that don't look like it. We may then ignore a competing product that may otherwise have strong bottom-up saliency. It is for this reason that – in one of the most famous psychology experiments of recent years[5] – most people fail to see a man in a gorilla suit amongst a group of people passing basketballs back and forth. The viewer is told to concentrate on counting the number of passes of the ball between those people wearing black T-shirts.

The task is fairly difficult and requires visual focusing, therefore elements that would otherwise pop out on our saliency maps – ie the man in the gorilla suit – get suppressed.

Indeed, similar to the *mere exposure effect* (described in Chapter 3), some researchers have coined the term *mere selection effect* to describe the finding that, if someone focuses their gaze on a product whilst simultaneously ignoring (or visually suppressing) other products, they are subsequently more likely to choose it.[6]

Designs or design elements can be thought of as having high or low saliency, but it depends on their surrounding context. For example, a bright-red pack design will not pop out on our saliency map if it is on a shelf next to many other competitive packs that are also red. Red may often be an eye-catching colour, but if you are just one red pack amongst many others, you are not as distinctive.

For this reason, just using a design that is bright, moving or unusual does not mean it will necessarily become salient and noticed. It all depends on the viewers' saliency maps. However, as long as the context and goals of the viewer can be understood and taken into account, the visual saliency of a design can be a powerful and useful concept.

Implied motion

Movement is one of the visual features that grab our attention quickly. Our ancestors would have needed to be aware quickly of anything moving, in case it was a predator. However, how can you use movement if you are creating still images? Implied motion is the neuro-design term for images that create the feeling of movement. Images of things that are obviously moving, such as a drink being poured from a bottle, can create the implied motion effect. Making an image lean forwards helps to non-consciously convey speed. When people walk quickly or run they may lean forward, but this artistic convention also extends to things like cartoons of cars moving forward quickly. This even extends to words. When asked to recognize names of animals that can move quickly, people were faster at recognizing fast animals when their names were italicized (ie leaning forward) such as *cheetah*.[7] Even arrows can evoke implied motion. Neuroscientists at the Salk Institute for Biological Studies in California discovered that images of arrows can activate neurons that are primed to detect motion – even though the arrows are just still images, and not themselves moving.[8]

The power of visual saliency

We have already seen how people often use mental shortcuts or heuristics to make decisions when they are unwilling or unable to make fully rational choices by logically weighing up all the relevant information. For example, in a supermarket, when shoppers are buying commodity goods that they are not hugely invested in, and are often distracted and under time pressure, the more a shopper looks at any particular product, the more likely they will choose it.[9] This effect is also independent of the shopper's own food preferences and how 'attractive' the pack design is (in an aesthetic sense).[10] Shoppers typically do not have time to examine every single product for an equal length of time, so what their saliency maps suggest they look at can influence their purchase considerations.

In one series of studies, researchers investigated the effect of visual saliency on choice of food products.[11] Participants were first asked to rate their personal food preferences on a 1–15 scale, and were then shown a series of images of food items and for each image were asked to select which item they would most like. The researchers boosted the visual saliency of one of the items on each image by making it brighter. The images were also exposed for different lengths of time, between 70 and 1,500 milliseconds.

There was a significant effect on the participants' choices of making the items more salient, particularly when the images were shown at the fastest speed. The researchers then repeated the study, but this time increased the cognitive load of the participants by giving them an extra task to do whilst they were making their choices. This was to replicate the fact that many shoppers are typically distracted (eg by talking to someone, or thinking about other things) whilst they are making their choices in the real world. In this version, the impact of visual saliency was even stronger. In fact, when shown the images for half a second, the effect of saliency was twice as strong as a one-point extra preference rating for the food. The effect of saliency was still significant when the images were shown at the longest duration (1,500 milliseconds).

These experiments show that when people are making quick choices, or choices whilst they are somewhat distracted, making a design more visually salient can make it more likely to be chosen. Whilst most experiments look at choices made under fast speed conditions, there is also evidence that the effects of more salient packaging can occur after a larger number of eye fixations around the shelves.[12]

Saliency mapping software

A number of different groups of academic neuroscientists have developed software algorithms that mimic the saliency mapping of humans. They are based on a number of sources of information, ranging from direct brain recordings from primates to eye-tracking studies on humans.[13] Saliency mapping software takes a digital image or video and analyses it pixel by pixel, taking into account the surrounding pixels. It looks for low-level visual features such as colour, brightness and orientation of lines, and produces a topographic map showing the most visually salient areas. Within seconds the user can have a heat map that overlays on the image zones of hot or cold colours, illustrating those areas that are likely to grab the most or least visually salient-driven attention. The opposite format output, a fog map, instead covers up all the non-salient areas in a misty fog, making only the more salient areas visible.

These computer algorithm-based maps have then sometimes been validated against eye-tracking data of where people actually look.[14] Researchers can subsequently improve the algorithm to up- or down-weight the importance of colour, brightness and orientation in order to mirror certain biases that humans have towards particular areas of their visual field (eg biases towards the upper or central areas of the visual field), or to reflect the tendency for people to look at text, or faces (both human and animal faces).[15]

Much of the impetus behind the development of these algorithms is to help computers to understand the visual world better, just as we as humans can. For example, enabling things like automatic detection of tumours in scan images, helping robots to navigate their surroundings, and automated centring and cropping of images around the most important object in the image.

Some of these algorithms are now available commercially to designers and marketers. They offer a more efficient way of rapidly testing the likely visual impact of designs than A/B testing on human participants.

In a similar experiment to the one that manipulated the visual saliency of food items, participants were presented with a series of photos of shop shelf displays of snack food items.[16] Before the test they rated each item that would appear on a scale, depending on how much they would like to eat it (and they were asked not to eat for three hours before the study, to ensure they were hungry and motivated!). Again, the photos were shown relatively quickly (between a quarter of a second and just over three seconds). After looking at each shelf display (containing 28 snack items) they had to choose which item they would want. Whilst they were doing this, their eye movements were monitored using an eye-tracking camera. The saliency of the items was not artificially tinkered with, but instead was measured using a saliency mapping algorithm. First, from the eye tracking they found that for every second a participant looked at one item, it increased by 20 per cent the chance that they would subsequently select that item. Perhaps not surprising, as one might expect shoppers' eyes to linger longer on things they are considering buying. However, they also found that where people first looked depended more on how salient the item was than how much they liked it. They also found that saliency affected where people looked, up to the seventh fixation. Whilst the greatest effect on choice was whether the person liked the item, it was also important whether it was salient. In other words: given the choice between several items that a person likes, they will tend to pick the one that is more visually salient.

Using saliency mapping software

Using saliency mapping software is cheaper and quicker than running eye-tracking studies. Therefore it is more practical for testing quickly and iteratively. For example, designers can check the effects that changes in different elements within their design – such as colour, shape, size, etc – make to its saliency.

Typical outputs include heat maps (colour overlays on an image, depicting areas of higher saliency with hotter colours), or their reverse: fog maps (obscuring the less salient areas, and keeping the more salient areas more visible). Fog maps are often easier and more intuitive to work with, and enable you to see clearly the areas with saliency (heat maps, by overlaying with colours, both obscure the image below, and can be confusing if the colours of the map match the colours that are naturally in the image). Other outputs include numerical scores for such things as the overall percentage of an image that is salient, or for more specific areas or objects within the image.

There are several potential ways to use visual saliency mapping software. First, to test the comparative saliency of several different design options to see if any are clearly better at standing out. Second, to test iteratively to maximize the visual impact of a design. For example, if you want to make a particular element of a design more likely to stand out, you can test the effect of changing things like colours, outlines or font sizes. Lastly, visual saliency software can be useful for testing the likely visual impact of a design in-context. For example, a pack design can be embedded in a shelf display image, an online ad on a webpage, or a print ad within a magazine. The only watch-out here is that the algorithm is literal: it will analyse the exact details of the image put through it. However, shelf displays, webpages and magazines can have a lot of different variety, so it may be worth producing several different versions of the context background (shelf, webpage, magazine) to hone in on the likely typical saliency effects.

Saliency mapping on webpages

The effects of making items visually salient are particularly relevant and useful to web design, especially given the importance of visual saliency when people are impatient or distracted (which web users often are). One study used a visual saliency model (based on intensity and colour contrast, size of an item, and how close it was to the centre of the page) to accurately predict where users will look on a webpage.[17]

The concept of saliency mapping can be used to design pages that guide users' attention to the things you most want them to look at, or to ensure that lower-priority elements are not grabbing too much attention and stealing it away from the areas where you would prefer they looked. It is possible to have a design that is aesthetically beautiful, but has low visual saliency and hence is likely to be overlooked.

Whilst visual saliency testing is cheaper and quicker than running an actual eye-tracking study, it should be borne in mind that there will also be top-down influences on where users look. For this reason, whilst visual saliency is a more accessible technique than eye tracking, it should not be seen as a complete substitute for it. As mentioned, things like food, drink or human faces can grab people's interest (even if it is not technically as fast as the bottom-up, *pre-attentive* processing I mentioned earlier in the chapter). Equally, there is often a central attention bias: when people are looking at displays or screens, they gravitate more towards items placed at the centre, and sometimes even within a design too (eg the centre of a pack design will get more attention than its periphery). For example, one eye-tracking study on people looking at just over 1,000 images of things like landscapes and portrait photos found that 40 per cent of eye fixations were within the central 11 per cent of the images.[18]

Web users also have expectations for where certain things should appear on a page, which have grown out of conventions in web design. For example, users expect to find the menu at the top of the screen. Expectations like this can also drive where users will look, given what their particular search goals are, and make things quicker to find.[19] However, items on a webpage that may not always be positioned in the same place (and hence users don't intuitively learn where to find them) can benefit from being made more visually salient.

A user's visual *entry point* – where they will begin their search on a webpage – is influenced by their mapping of what is visually salient on the page.[20] This means that visual saliency can also help solve usability issues on a site. Finding a menu, or the key information that will allow users to navigate a page, can be made easier by selectively making key design elements or text more salient.

One consideration with saliency on webpages is whether to make banner ads salient. Ever since the early days of the web, researchers have noticed that banner ads tend to get ignored and have low click-through rates (a phenomenon known as banner blindness). The theory is that users learn to identify things that look like ads, even in their peripheral vision, and ignore them. Is there a danger of making banner ads too salient? If users already have a tendency to quickly notice when a design is an ad, could making it salient simply speed up this recognition and enhance banner blindness? Indeed, could making them more salient also harm the remainder of the designs on the page, and potentially annoy users by attempting to drag their attention on to something they are resisting looking at, and away from things they want to look at? However, initial research looking at making banner ads more salient seems to show that this is not a problem. Nevertheless, as little is yet known about the possible interaction of banner blindness with visual saliency, it may be worth testing the effects of making your banners more or less salient. Similarly, using animation can increase a user's cognitive load and cause visual confusion, particularly if it is within an already visually complex environment.[21]

How designers can use visual saliency

Any design that needs to be noticed in a competitive environment, that needs to arrest attention, or that needs to draw attention to particular elements within the design can benefit from considering its visual saliency. For example, the colour properties of a photo could be adjusted to make it more visually salient. When choosing the colour scheme of a design that needs to grab attention, more contrasting colours should be used; or items on a webpage that need to stand out strongly could have an element of animation (such as pulsing or flickering effects).

Designers can experiment with running their designs through a visual saliency algorithm, but they can also just use the theory of visual saliency. In particular they should bear in mind the following 'golden three' elements when making a design element more salient:

- **Colours:** brighter, more luminescent colours, and contrasting colours within a design, are more likely to be visually salient. In particular, the more edges of contrasting colour within a design, the more salient it is likely to be.
- **Size:** larger objects in a design (perhaps obviously) are more likely to be seen, but size also affects things like text: thinner, smaller fonts will be less salient than larger, thicker fonts.
- **Patterns:** patterns of objects that stand out against their surroundings, shapes that are unique compared to those around them, or areas of contrast where one shape overlaps another are highly salient.

Even when top-down saliency is likely to be important, maximizing the saliency of bottom-up image features can still be important. For example, if you know that your customers are searching for something by a particular element of its design (eg a colour, or a particular logo), you can make that element more salient and hence help make it easier to find.

Visual saliency can also be particularly useful in developing packaging designs. As shown in Figure 6.3, showing how well a design stands out against the competition can help it to stand out against the competitor packs. The visual-saliency fog maps cover up the less salient packs and reveal those that are more salient.

Figure 6.3 Simulation of pack designs on-shelf, and a saliency fog map showing how well each pack stands out

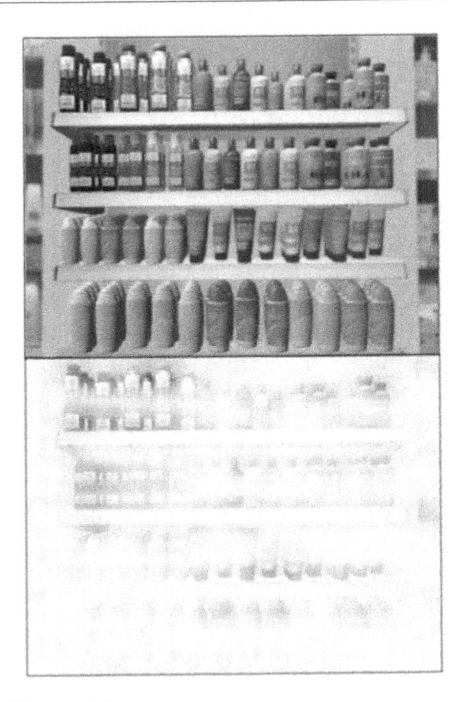

SOURCE: Image created by Paravizion.com

One consideration with using visual saliency is to balance the need for making a design salient with the benefits of it being prototypical (see Chapters 3 and 4 for more on prototypicality). For example, on a webpage, consider where users expect to find items; if you cannot place them in areas where they might be expected then consider making them more salient.

It is not necessarily the overall saliency of a design that counts, but whether it is focused. For example, if a design has too many highly salient elements that are positioned away from each other, it could make the design too confusing and complicated, and viewers might not know where to look. Focused saliency over a smaller number of important areas is better than diffuse saliency over many.

Finally, technically things that are 'visually salient' – in the neuroscience sense of the word – are things that stand out due to their basic visual properties. However, there are top-down elements that are also likely – in general – to grab attention quickly. For example, faces – of both humans and animals, particularly faces displaying emotional expressions. Also design elements that create a frame. We are used to frames signalling that they contain information we should look at inside – whether they are TVs, paintings, mirrors, or the outer frame of a computer screen. Placing a frame border around text or design elements can be an effective way to focus attention inside it.

Summary

- Our visual attention is guided by both bottom-up (the visual features of what we are seeing) and top-down (our memories, expectations and search goals) processes.
- As we can only process details in one area at a time, our visual system needs a way of prioritizing what to look at first. To do this it automatically computes a saliency map of things that stand out from their surroundings.
- Bottom-up features of an image can make it visually salient, although there can also be an influence from top-down processing. Bottom-up processing is essentially the same for all humans, so what is visually salient from a bottom-up perspective will be salient to everyone.
- Based on a number of sources of neuroscience research, neuroscientists have created visual-saliency mapping software that can take any digital image or video and predict within seconds which elements of it will grab attention.
- Items in-store that are visually salient get looked at sooner, for longer, and are more likely to be chosen.

Notes

1 For example, see: http://everythingwdisneyworld.com/2012/07/go-away-green-and-no-see-um-gray-at/ (last accessed 25 August 2016).

2 Milosavljevic, M and Cerf, M (2008) First attention then intention: insights from computational neuroscience of vision, *International Journal of Advertising*, 27 (3), pp 381–98.

3 Mormann, MM, Towal, RB and Koch, C (2015) Visual importance of marketing stimuli, *Insights from Visual and Computational Neuroscience*, 29 December.

4 Wolfe, JM and Horowitz, TS (2004) What attributes guide the deployment of visual attention and how do they do it?, *Nature Reviews Neuroscience*, 5 (6), pp 495–501.

5 Simons, DJ and Chabris, CF (1999) Gorillas in our midst: sustained inattentional blindness for dynamic events, *Perception*, 28 (9), pp 1059–74.

6 Janiszewski, C, Kuo, A and Tavassoli, NT (2013) The influence of selective attention and inattention to products on subsequent choice, *Journal of Consumer Research*, **39** (6), pp 1258–74.

7 Walker, P (2015) Depicting visual motion in still images: forward leaning and a left to right bias for lateral movement, *Perception*, **44** (2), pp 111–28.

8 Schlack, A and Albright, TD (2007) Remembering visual motion: neural correlates of associative plasticity and motion recall in cortical area MT, *Neuron*, **53** (6), pp 881–90.

9 Krajbich, I, Armel, C and Rangel, A (2010) Visual fixations and the computation and comparison of value in simple choice, *Nature Neuroscience*, **13** (10), pp 1292–98. See also: Clement, J, Aastrup, J and Forsberg, SC (2015) Decisive visual saliency and consumers' in-store decisions, *Journal of Retailing and Consumer Services*, **22**, pp 187–94; and Isham, EA and Geng, JJ (2013) Looking time predicts choice but not aesthetic value, *PloS One*, **8** (8), p 71698.

10 Mormann, MM, Navalpakkam, V, Koch, C and Rangel, A (2012) Relative visual saliency differences induce sizable bias in consumer choice, *Journal of Consumer Psychology*, **22** (1).

11 Orquin, JL and Lagerkvist, CJ (2015) Effects of salience are both short-and long-lived, *Acta Psychologica*, **160**, pp 69–76.

12 Fecteau, JH and Munoz, DP (2006) Salience, relevance, and firing: a priority map for target selection, *Trends in Cognitive Sciences*, **10** (8), pp 382–90.

13 Dupont, L, Ooms, K, Antrop, M and Van Eetvelde, V (2016) Comparing saliency maps and eye-tracking focus maps: the potential use in visual impact assessment based on landscape photographs, *Landscape and Urban Planning*, **148**, pp 17–26.

14 Judd, T, Ehinger, K, Durand, F and Torralba, A (2009) Learning to predict where humans look, in *Computer Vision, 2009 IEEE 12th International* Conference, pp 2106–13).

15 One watch-out with correlating bottom-up visual salience models with where users look, is that often people will look at an object within a scene because they are interested in it, but it will also be salient within the software model because the edges of the object stand out. This can make it look like the person looked at it because it is salient, when that was not the real cause, eg see: Carmi, R and Itti, L (2006) Visual causes versus correlates of attentional selection in dynamic scenes, *Vision Research*, **46** (26), pp 4333–45.

16 Mormann, MM, Navalpakkam, V, Koch, C and Rangel, A (2012) Relative visual saliency differences induce sizable bias in consumer choice, *Journal of Consumer Psychology*, **22** (1).

17 Jana, A and Bhattacharya, S (2015) Design and validation of an attention model of web page users, *Advances in Human–Computer Interaction*, 373419:1-373419:14, p 1.

18 Roth, SP, Tuch, AN, Mekler, ED, Bargas-Avila, JA and Opwis, K (2013) Location matters, especially for non-salient features – an eye-tracking study on the effects of web object placement on different types of websites, *International Journal of Human–Computer Studies*, **71** (3), pp 228–35.

19 Shen, C and Zhao, Q (2014) Webpage saliency, in *Computer Vision – ECCV 2014*, Springer International Publishing, pp 33–46.

20 Still, JD and Masciocchi, CM (2010) A saliency model predicts fixations in web interfaces, in *5th International Workshop on Model Driven Development of Advanced User Interfaces (MDDAUI 2010)*, (April), p 25.

21 Breuer, C and Rumpf, C (2015) The impact of color and animation on sports viewers' attention to televised sponsorship signage, *Journal of Sport Management*, **29** (2), pp 170–83.

07 Visual persuasion and behavioural economics

Figure 7.1 By paying attention to how people naturally behave, and the shortcuts they take, designers can create more persuasive designs

Several years ago, illusionist Derren Brown performed an interesting trick on his TV show. He brought two advertising creatives to an office and gave them half an hour to drawn a poster ad for a fictional animal taxidermy company. They were told that their design should contain a company name, a logo and a strapline. Being a mentalist magician, before they began he placed an envelope containing his prediction of their ad on the desk in front of them, unopened.

When the creatives revealed their design to him, it was almost exactly the same as the sketch he had previously made, predicting theirs. It contained big zoo gates, designed to look similar to the gates of heaven, a bear sitting on a cloud playing a harp, the name 'creature heaven' (the ad creatives called theirs 'animal heaven'), with angel wings over the name, with the strap line 'Where the best dead animals go' (their line read: 'the best place for dead animals'). They thought they had produced a truly creative, original design, yet Brown had predicted almost exactly what they would draw.

Brown's explanation for how the trick worked was as follows. The taxi ride that the creatives made on their way to the film set-up was carefully contrived to expose them to certain visual elements. As the taxi drove down the road, Brown had prepared a series of visuals that they would see. First, their ride took them past the gates of London Zoo. At a crossing, the taxi stopped whilst a group of children crossed the road, each of them wearing jumpers with a picture of a zoo gate. They drove past a window that had some posters stuck to it, two of which said 'where the best dead animals go', and also a chalkboard with a picture of two wings, in a symmetrical pair, over the words 'creature heaven'. They also drove past a shop displaying a large harp in its window. Finally, as Brown gave them their task, there was a giant bear model behind him. In other words, the creatives had been carefully visually primed along their cab route with a series of images and words, placing these images into their minds so that they would be the very ones that bubbled up into their imaginations when it came to designing the poster. When they were creating their poster, they were unaware that these ideas had been deliberately placed in their minds.

What does this tell us about how people are affected by images in real life? Of course, him being an illusionist, and it being a TV show, it is impossible to know for certain whether we can trust Brown's own explanation of how his trick worked. Also, the situation was somewhat contrived, with a degree of control over what the creatives saw and were asked to do that was more extreme than typical everyday situations. However, the idea that we can be influenced by images, even without consciously being aware of it, has got some scientific evidence behind it.

Priming

Subliminal advertising is the urban myth that TV ads flash up words and images – too fast for us to consciously notice – in order to influence viewers almost hypnotically to buy their products. In the 1950s and 1960s, scary stories about subliminal advertising came to the public's attention through books such as *The Hidden Persuaders* by Vance Packard. The origins of many of these scary stories turned out to be hoaxes, but the notion stimulated the public's fears.

By 1974 a report by the United Nations declared that subliminal advertising was a 'major threat to human rights'.[1] Despite there being no evidence for its existence, the practice was outlawed. Subliminal advertising per se might not be used, yet the things that we see without paying too much conscious attention to can often influence us.

Our non-conscious minds are optimized to give us quick responses in the moment. When we are making decisions, particularly relatively unimportant decisions, we are more likely to consider things that are either within our immediate environment or that easily come to mind. We don't search for extra information. This is how priming can affect our behaviour. Priming, as we saw in Chapter 3, is a little like our non-conscious System 1 brain acting as a personal assistant, pulling out all the relevant files for an executive to make a decision; but what is deemed relevant is heavily biased towards things we have just encountered. Studies have shown that when people are primed with a concept – like 'rudeness', for example – they are then more likely to interrupt someone, and when primed with the concept of old age, they are more likely to walk more slowly.[2] Similarly, people primed with the concepts of luxury or thrift subsequently make more indulgent or economical spending choices respectively.[3] In common with Derren Brown's TV stunt, these people were influenced by ideas without them being consciously aware of it. However, whereas Brown's trick was more literal (ie see an image of a set of gates leading to drawing a set of gates), these priming studies show it can also be more general. General concepts, such as rudeness, old age or indulgence, were triggering related concepts in the minds of participants and affecting their behaviour. This shows how priming also triggers the things that we mentally closely associate with the prime: such as old age being associated with walking more slowly.

The images we see immediately before or at the time of a decision can influence our behaviour without having to persuade us rationally.

In Robert Cialdini's book *Pre-Suasion* (2016) he gives a good example of how context can prime us to be receptive to different types of advertising messages. He realized that two of our most powerful drives – to stay safe and to find a partner and procreate – can lead to two opposing behaviour strategies at any particular moment: the desire to be supported by fitting in with the crowd (safety in numbers) or to stand out from the crowd (in order to appear more attractive to a prospective partner). If this were true, then it would mean that if we felt scared we might be primed to be more receptive to messages that emphasized melding with the crowd. Conversely, if we are in the mood for romance we might be more receptive to messages that emphasized being different from the crowd. He ran an experiment and found exactly this. When people had been primed to be scared, by viewing a violent movie, people responded more favourably to a museum ad that emphasized fitting in with the crowd: telling people that the museum had over 1 million visitors per year. However, when they had been watching a romantic movie the same ad was ineffective, whilst one emphasizing standing out from the crowd was effective (and ineffective for those who watched the fear-inducing movie).[4]

The subtext of things can also be just as important to us as the overt message. For example, we are used to paying attention to a person's tone of voice, facial expressions and body language when they are talking to us. A long pause before someone answers a question can sometimes communicate more than the actual content of what they say. Whenever a message is conveyed, we don't just decode the message, but we often question the format of the message or why it is being said. For example, long text copy on a webpage might contain a lot of information that the person does not find strictly necessary, but the fact that you have gone to the effort to write it can communicate reassurance to them.

Persuasion is not always conscious

One of the earliest models of how advertising persuades people (dating back to the 19th century) was AIDA, an acronym for attention, interest, desire and action. In other words, first you need to grab attention, then once someone is attending to your ad you need to generate interest. Their interest in your message must then blossom into desire and, finally, after their desire is stimulated, they will take action and buy your product. Despite it involving emotion (ie desire), it is a very rational model of how people behave. We now have a more sophisticated understanding of people, and the non-conscious mind. AIDA is a fairly conscious model: at each of the AIDA steps, a person could, in theory, tell you which step they were at. However, as we have seen in earlier chapters, there are other ways in which persuasion can happen:

- **Saliency mapping:** this is slightly closer to AIDA in the sense that it involves grabbing attention. However, the factors that are involved – the low-level design features such as colours, contrasts, brightness etc – are largely operating below a person's conscious awareness, and there is little rational persuasion going on.
- **Processing fluency:** people may choose something simply because it has a more simple, easy-to-process design.
- **First impressions:** people's evaluation of a design can be made very quickly, in less than a second. This is clearly not conscious, rational persuasion as we know it!
- **The affect heuristic:** simply making something emotionally charged could sway people in its favour. Not very rational. Also, colours or the different sensory associations that someone has with your message can trigger desire, below the level of conscious awareness.

Furthermore, psychologists now know that persuasion, in the sense of attitude change, is not always needed in order for someone to take action. The old model is that someone has an attitude or set of beliefs about your brand, product or service. They might be favourable, in which case their buying behaviour will fall into line with their attitude and they will take action accordingly. Or their attitude will be unfavourable, in which case they need to be consciously persuaded in order that they will change their behaviour. However, in many instances we now know that often behaviour comes first, and attitude follows. A person may act because something – such as their first impressions or the processing fluency of the design – has non-consciously triggered their behaviour, and then they later bring their attitudes into line with their behaviour. We like to feel that we are in conscious control of our behaviour, and we like to be consistent. Hence there are strong pressures for our attitudes to fall into line with how we have acted, so that our own behaviour makes sense to us. When our thoughts are not consistent with how we have behaved, this triggers a tension or discomfort that we seek to lessen. Psychologists call this *cognitive dissonance*.

In particular, strong attitudes are more predictive of behaviour, whereas those with weak attitudes can have their attitudes driven more by their behaviour. In one study, researchers measured participants' attitudes towards the charity Greenpeace. A week later the participants were asked to give a donation. The donations received were very predictable for those with strong attitudes, but not for those with weak attitudes. But if those with weak attitudes did donate, it shifted their attitudes.[5] When people have a weak attitude towards something it is harder for them to remember their attitudes clearly, so they are more likely to construct them on the spot, meaning that their behaviour has a chance to bias their attitude.

This is similar to how our behaviour can drive our moods. If you are feeling down, emotionally, going for a walk or taking exercise can make you feel better. As the psychologist William James, writing in the 1890s, put it: 'Sit all day in a moping posture, sigh, and reply to everything with a dismal voice, and your melancholy lingers... If we wish to conquer undesirable emotional tendencies in ourselves, we must... go through the outward movements of those contrary dispositions which we prefer to cultivate.'[6] Given this, we can see that trying to predict people's behaviour from their attitudes, or conducting attitude questionnaires, may be an inadequate approach.

Attitudes are predispositions towards things: we are positively or negatively inclined towards certain brands or products, for example. Much of marketing thought has assumed that these attitudes are the product of rational thinking. For example, in order to get a person to be positively inclined towards your product you need to persuade them of its benefits. However, in most developed markets now, many products are functionally the same as their competitors. Differences in sales cannot usually be accounted for by one product being functionally superior to its alternatives. Here is where design comes in: we have seen the power of design to affect people. Optimizing a product's design can be enough to differentiate it, even if its basic qualities aren't that different from the competition. For example, making it more visually salient, easy-to-process, and triggering the right sensory and emotional associations.

Another misleading feature of the AIDA way of thinking is that people often think of persuasion along the lines of 'moving' something in the real world. If someone's attitude needs to change a lot in order to change their behaviour significantly, then a lot of effort and energy needs to be applied. Just as if you had to move a huge rock, you would need to apply a lot of energy to the task. This can mislead marketers into strategies that may be unnecessarily complex and costly.

For example, imagine you are commissioned by an environmental charity to help increase 'green' behaviours such as recycling. If approached from an AIDA perspective you would try to change people's attitudes. You might seek to convince them of the environmental impact of not behaving in a green way. You might try to explain to them the results of scientific studies, or of models and projections of future environmental degradation if populations do not change their behaviours. However, a smarter approach might simply be to look for ways to change their behaviour more directly, such as making recycling bins easier to find by making them more visually salient.

Take the example of pensions. In developed countries many people are not saving as much as they could afford to for their retirement. The traditional AIDA solution to the problem might look like this: we need to shift people's attitudes to convince them to save more. Therefore we need to show them the facts and figures. Graphs would be good. Explain the statistics, the investment plans, model how much the person needs to save, and so on. However, looking at the problem through the lens of our understanding of the non-conscious mind we might take a radically different approach. It involves some concepts I have not introduced yet, so I would like to pause the question there, and we will return to it shortly and see if you can devise a better strategy!

Liking versus wanting

The measurement of beauty in the brain raises an important consideration: not everything we like to look at is necessarily beautiful. We can recognize and admire high levels of artistic merit in an image without finding the image itself beautiful. Similarly, we can enjoy looking at something without wanting to possess it. Psychologists call this *disinterested interest*.

Our brains possess two separate systems for liking versus wanting something we see. Each system involves distinct neural circuits and neurotransmitter chemicals. For example, the wanting system uses dopamine; the liking system involves opioids and endocannabinoids. Dopamine is often mistakenly talked about as a pleasure chemical (as neuroscientists used to believe this back in the 1980s); in fact, its real purpose is to drive desire. The wanting system (desire or longing) is more easily triggered than the liking system (pleasure itself). Neuroscientist Dr Kent Berridge, the pioneering researcher behind discovering these two separate systems, says: 'It's easy to turn on intense wanting: massive, robust systems do it. They can come on with the pleasure, they can come on without the pleasure, they don't care. It's tricky to turn on the pleasure. This may explain why life's intense pleasures are less frequent and less sustained than intense desires.'[7]

▶

Whilst it is easy to ask someone if they like something, this does not necessarily mean it will translate into them buying it. For example, in one study a number of teenagers, whilst in an fMRI brain scanner having their brain activity monitored, listened to a series of new pop songs they had not heard before. They were also asked to rate how much they liked each track on a scale. After all the tracks had been released and sales data was available, researchers found that the brain activity was somewhat predictive of how many people actually bought each song – particularly for the tracks that were commercial failures – but the teenagers' 'liking' responses were not predictive.[8]

Behavioural economics: shortcuts to decision making

Behavioural economics is a relatively new field that applies understanding of psychology to the economic choices people make. The old, rational model was that people seek to maximize the benefits they get from spending their money and they do this by performing some form of mental calculation, taking into account the potential benefits they would get from each product, its cost, volume etc – almost as though consumers had an inner accountant in their brain. Whilst we do something like this sometimes, we usually don't have the time or mental energy to perform such an exhaustive analysis of the options available. Instead we use mental shortcuts, such as our gut feeling about the options open to us. Behavioural economics studies these shortcuts.

As mentioned earlier, marketers who have adopted the rational/AIDA way of thinking will often assume that they have to apply a lot of energy to logically persuade someone of the benefits of their product or service. However, behavioural economics, through recognizing that we often make decisions via quick mental shortcuts, can often provide solutions that are relatively quick and simple.

Often the key to behaviour change can be simply removing barriers. We often avoid things just because they take too much effort. Not only physical effort (ie going to the gym), but mental effort. There are too many sub-decisions that need to be made to arrive at the main decision. Or too many forms to fill in. Too many steps.

There are three main types of barrier that are usually blocking people from buying something:

1 Risk
   Evolution has bequeathed our brains with a particular aversion to risk. For our ancestors, risk could mean losing their lives. The costs of not avoiding risk loomed large. We are sensitive to risk, and even if we want something, if there is any risk associated with it then we might just pass.
2 Uncertainty
   If there is even a small amount of uncertainty associated with buying something, it can prevent people doing so. If someone sends money, are they going to get a receipt, confirmation? Will they know how long it will take for their order to be fulfilled?
3 Difficulty
   As we saw in Chapter 3, making designs easy to understand is a powerful technique. People will often shy away from tasks that seem to require a lot of effortful thought, whether those tasks are difficult, or merely require multiple stages. For example, removing the difficulty of making an order or filling in a form is often all that is needed to 'persuade' a new customer.

   'I have read and understood the terms and conditions' may be the most frequently told lie in the world today! Often long passages of legal text are presented on websites, which in practice are unreasonable for users to read through and understand. Of course they need to be there for legal reasons, but they seem to me to be an example of where companies follow the letter of the law without following the laws of how people think.

   Designing forms is a particularly rich area for improving the effectiveness of websites. Most people do not enjoy filling in forms: it's tedious, and people also can have varying levels of resistance to giving out their personal information (eg because they have privacy concerns, or they wonder whether they will get lots of spam e-mails or texts).

Another useful barrier to dissolve is guilt. With so many people on tight budgets, any discretionary spending – such as on small treats – can cause an accompanying feeling of guilt. Are there ways in which you can lessen the guilt?

One of the main areas of interest in behavioural economics is uncovering the non-conscious shortcuts or rules of thumb that people use to make decisions. These are called heuristics.

Examples of heuristics

Mental availability

Similar to that which we saw earlier with priming, when we are making choices or decisions, the information that is available to us in the moment has a disproportionate sway over us: for example, what we can see around us, or the information that easily pops into our minds. Whilst some people count the calories they consume, most of us don't know how many calories we take in daily or weekly. So this information is not available to us when we are making our choices. We could find out this information if we put the effort in, but typically people don't. However, if a supermarket checkout or online grocery basket totalled up the number of calories (and also perhaps the fat and sugar content) in our weekly shop and told us, suddenly this information would be available and could potentially affect our behaviour (ie we may then decide to make more healthy food choices).

The availability heuristic is one of the reasons why we are so bad at judging and acting on probabilities. For example, people often fear dying in a plane crash, or a disaster of some kind, more than they fear dying of heart disease, even though heart disease is statistically more likely. This is because images of things like plane crashes come more easily to mind than images of heart disease. Heart disease is more abstract, it doesn't naturally translate into a clear mental picture. The more mentally available an idea is, the more likely an outcome it can seem.

Many countries – such as Australia, France and the UK – have now adopted strict controls over how cigarettes can be sold: for example, not allowing them to be visible in-store (being hidden behind screens), and even removing all branding and design from the packets themselves. This works to limit or even destroy the mental availability of the cigarette products. It is easier to recognize information than recall it. Simply seeing a brand design displayed is a reminder of it, which is far easier than having to search our memories to recall it and ask for it at the counter. The design elements on a pack of cigarettes also act as memory 'hooks': multiple ways for us to remember a brand, be it by colours, shapes or the typeface of the brand name. By removing these, it further acts to undermine our ability to recall the brand.

There is a role for design in helping to increase the mental availability of concepts that you want your potential customers to think about. For example:

- If you are trying to communicate an abstract idea, think about how it can be visualized. This might be literally showing an image of it that is not often seen. For example, in the case of heart disease it might be showing what clogged arteries actually look like. Things like cutaways that reveal the inner workings of things can be good here. Or if the concept is truly abstract, can you devise a good infographic to explain it? Or can you think of a striking visual metaphor that helps drive home the concept?

- Does your brand, product or service have images that come to mind? Is it visually memorable?
- Can you use multisensory integration (see Chapter 5) to help make your brand's distinctive assets more memorable?
- Feeding back customers' own data can make information available to their choices, and hence affect their decisions. For example, information on how much they have saved shopping with you in the past year, or information on how much they use a particular service.

Anchoring and framing

When we are making choices, we usually have a decision or choice set: the alternatives we are considering. Some of the alternatives may be dismissed quickly, for example if they are too expensive for us. Yet their very presence can shift our perceptions. As market traders have known for millennia, showing a customer a more expensive product choice can make the first option seem more reasonable:

- Provide relevant, personalized choice comparisons that show what good value your product/service is.
- Can you visually depict your product within a choice set that makes it seem like a good option?
- If you are one of the more expensive options within your category, can you show how your product/service is an alternative to something in a different category that is typically more expensive? For example, a piece of exercise equipment or a health-tracker wearable could be compared to paying for a gym membership and personal trainer.

Hyperbolic discounting and loss aversion

Hyperbolic discounting simply means that we tend to value pleasures and rewards in the moment more than in the future. We also tend to value things more once we own them.

If we buy a pair of shoes that don't fit, but we have lost the receipt, we may feel uncomfortable with just throwing them away. It feels equivalent to throwing away the same amount of cash that we spent on the shoes. We might just be happier to leave them in the wardrobe unworn. Fast-forward several years and we rediscover the shoes in our wardrobe. They are now old. Their resale value has gone down, so throwing them away now doesn't feel so painful. Yet, this is not very rational if we think about it. The outcome in both cases is the same, but one option is more comfortable to us than the other.

We are also more sensitive to the possibility of loss rather than opportunity of gain. Loss aversion is the reason why free trial schemes are often so effective. If people have the experience of being a member of your service or owning your product (albeit temporarily), they are more likely to value it than if they had never had access or ownership. Hence they are more likely to want to continue owning it.

Related to loss aversion is the fact that we are more sensitive to negative than positive concepts. Professor Nilli Lavie, a psychologist at University College London, says: 'We cannot afford to wait for our consciousness to kick in if we see someone running towards us with a knife or if we drive under rainy or foggy weather conditions and see a sign warning "danger". Negative words may have more of a rapid impact – "Kill Your Speed" should work better than "Slow Down".'[9]

Calculate the risks of loss for the person if they don't buy your product/service, and communicate it to them.

Social proof

If we see other people doing something, it can immediately lessen our worries about its risks and uncertainties. Following others can be a shortcut to making a decision. Being part of the crowd is a strategy that we have used since early in our evolutionary history to make us feel safer.

Letting people know about how others like them have used your product or service can help provide social proof, alternatively, feedback information about other people who are similar to the customer. This might be something like telling them how many satisfied customers in their city you have served in the last year, or about additional products and services that people like them have found useful.

Fairness and reciprocity

Not necessarily strictly a heuristic, but a feature of how we learn to interact with others, is that we expect our interactions to be fair and balanced. As our interactions with a company over time can start to feel like a relationship, we carry some of the same expectations for it as with a personal relationship. If we have been loyal to a particular brand or company we can become particularly angered if they treat us unfairly. It doesn't feel right to us that we should be treated the same as a new or casual customer.

Revisiting the pension challenge

Now that we have looked at some of the behavioural economics techniques for making choices easier for people, take a moment and think back to the pension challenge I described earlier. Think about the following features of most pension services:

1. They involve us paying money now for benefits we will not experience for a long time.
2. Almost every element of dealing with the pension requires lots of form filling and reading of often complex legal and financial text.
3. If you make a payment into your pension it does not usually give instant feedback to reassure you that the money has gone in. In other words there is an element of uncertainty.
4. Once you have set up a pension, making a one-off payment into it can be a complex process. Each time you want to do so, you need to remember that process: you have to retrieve your password, username or pension number, find the provider's phone number or account number to make an electronic transfer on internet banking, or – even worse – find their postal address in order to send a cheque.
5. Finally, aside from the paperwork that you probably keep filed away, your pension service is invisible to you in daily life. You don't see any visual reminders of it. It is out of sight and out of mind.

Considering these features, take a moment to think about how the principles of neuro design and behavioural economics could be used to better match our non-conscious preferences. My suggested solutions are set out below.

Make paying into a pension quick, fun and easy

Many of us are now accustomed to accessing services on our smartphones and we expect our apps to be simple and fun. Creating a pension smartphone app could be a good way to increase the ease of paying into a pension. Currently with most pensions people have to exert a considerable amount of effort (ie they have to hunt for their documents to find their pension details, the details for how to pay money in, then they might have to make a phone call, log in to their online bank, or even post a cheque). In contrast, a smartphone app can be set up to perform this with just one or two clicks. The other benefit of a smartphone app is that it is more accessible. We carry our smartphones with us wherever we are: they are never far from us. Seeing the icon of the app on the phone is a constant reminder of the service: it doesn't fall out of our awareness in the same way as do forms and paperwork stored in a drawer. Smartphone apps for pensions could also change our perception of paying into a pension. Currently the effort associated with paying money in means that we are more likely to do it only when we have a significant payment to make. The ease offered by smartphones could mean that we would be more inclined to make smaller and more frequent payments.

Apps also offer the possibility of instant feedback, confirming that our payment has been added, and always offering a quick way to check the current total of our pension account. The ability to provide this information quickly and easily helps to remove the pain of uncertainty.

Here are some other suggestions based on the concepts we saw in earlier chapters:

- **Processing fluency:**
  - Make sure that the legal and financial information is translated into an easy-to-read style (obviously within the legal constraints that pension providers have to abide by).
  - Use infographics to illustrate the financial models of the different pension options, eg risks and possible returns over time.
- **First impressions:**
  - Many people already have a range of negative associations with pensions, so distinguish your service quickly from the competition. The first contact points that people have with your pension service – whether in ads, on your website or from leaflets (eg such as those in banks) – should immediately convey the simplicity and ease of your service.
  - Most pension documents and websites look complex and boring. Make them look beautiful yet simple.
- **Visual saliency:**
  - Most pension websites and documents contain lots of information, yet some pieces of information are more important than others. Help users to understand which bits of information are of higher priority, or which they need to attend to most closely, by making them more visually salient.
- **Multisensory stimulation:**
  - Incorporate different senses in your communications with users. For example, if there was a smartphone app to enable quick payment into

a pension, it could show stimulating animations and sounds when you make a payment, and perhaps make the phone vibrate. (The hyperbolic discounting heuristic works against pensions: we are asked to sacrifice now – by investing some money in the present moment that we could be using for immediate benefits – in order to receive benefits in the future. By adding in even small benefits or rewards to the process it can help to lessen this barrier.)

Some pension providers may employ some of these techniques, but as at the time of writing I am not aware of any that have fully taken advantage of all of them. Of course, the chances are that you are not involved in marketing or designing pension schemes! However, the above is just an example of how neuro design and behavioural economics thinking could revolutionize how people interact with a service.

Visual nudges

Nudges are simple, quick techniques to influence or evoke behaviours. An important feature of a nudge is that it should not be coercive. It is not simply forcing behaviour by cutting off other choice options, or by promising punishment if the person doesn't comply. Instead it prompts or biases people to behave in the desired way.

Nudges can be visual. For example, toothpaste brand Colgate had found that traditional campaigns to remind children to brush their teeth after eating sugary foods were not particularly effective. Whilst the message may be understood when it is delivered, it's soon forgotten. It is not mentally available at the key moment when it needs to be (ie just after someone has eaten sugary food). Colgate worked with an ice-cream brand to create a clever visual nudge: the stick at the centre of the ice cream was designed to look like a toothbrush[10] – hence offering the reminder, in a fun way, at the critical moment when it was needed.

Driving is a good example of where visual nudges have been used. Driving can be a very System 1 activity: many of our actions when driving are almost automatic. Drivers have to think and respond fast, and the consequences of bad driving can be very costly. Drivers use road signs as quick System 1 cues: for example, speed limit signs. Interactive road signs that quickly display a smiling or frowning face according to whether you are driving within or above the speed limit have been shown to be highly effective in reducing drivers' speeds.

Traffic-light icons on food products are another good visual nudge. These turn nutritional information from numbers into colour-coded circles, like a road traffic light. Testing has shown that having four colours, and also using a human figure in the design, can make them more effective.[11]

Key concepts

Priming

The way that just seeing (or otherwise experiencing) something makes that information more available to our brain, and hence can influence our decisions and behaviour.

Liking versus wanting

Our brains have separate processes for liking versus wanting something we see. Our wanting system is more extensive and easily triggered.

Behavioural economics

The field that studies how we make economic decisions based on mental, System 1 shortcuts or heuristics.

Visual nudges

Imagery that primes or influences us, without us necessarily being aware of it, but without forcing us.

Creating/testing persuasive imagery and nudges

If you visit most parks you will see people walking along the pathways provided. However, often on the grass you will also see worn paths where the grass has been trodden down or worn away through people walking over it. These areas of wear are like user-generated paths.

This tells us two things. First, that the visual cue of the provided paths is a powerful signal to people where to walk, but second, that when the designed network of paths does not lead people where they want to go, people will make their own paths.

With behavioural economics style techniques and nudges it is important to test their effectiveness. Build the path and see if people walk on it. Online this is obviously a lot cheaper and quicker to do as you can perform A/B testing: some people see one image, some another, and you can measure which is more persuasive.

However, whilst you can use the heuristics described earlier as a guide to creating persuasive images and nudges (and more lists can be found online if you search for 'behavioural economics heuristics'), it can also be revealing to watch what your viewers or users like to do (where they walk their own paths in the grass), or to think about the circumstances in which users already perform the behaviour you would like to provoke. What visual triggers are usually present when they already perform that behaviour? Visual influence is often more about setting the scene for behaviours in this way than trying to persuade people rationally.

Summary

- Persuasion is not always conscious and rational. The AIDA model – which a lot of marketers still base their thinking on – is outdated and does not take into account the non-conscious shortcuts that can lead someone to being persuaded.
- The phenomenon of cognitive dissonance explains how our behaviour can sometimes influence attitudes (rather than vice versa, which is the traditional expectation).
- Behavioural economics is a field that uses understanding of psychology to model how people make economic choices, often via non-conscious mental shortcuts or heuristics.
- Rather than expend effort in trying to rationally persuade people, behavioural economics solutions often use 'nudges': simple triggers that provoke the desired behaviour.
- Some of the most important heuristics include: availability (the information immediately to hand has a disproportionate effect on choices and judgements); anchoring and framing (the comparisons people make bias their choices); loss aversion (we are more sensitive to fear of loss than opportunity of gain); social proof (looking to others to guide our behaviour); and reciprocity (feeling that it is fair to do something for others if they have done something for us).

Notes

1. Quoted in: http://news.bbc.co.uk/1/hi/health/8274773.stm (last accessed 25 August 2016).
2. Bargh, JA, Chen, M and Burrows, L (1996) Automaticity of social behavior: direct effects of trait construct and stereotype activation on action, *Journal of Personality and Social Psychology*, **71** (2), p 230.
3. Chartrand, TL, Huber, J, Shiv, B and Tanner, RJ (2008) Nonconscious goals and consumer choice, *Journal of Consumer Research*, **35** (2), pp 189–201.
4. Cialdini, R (2016) *Pre-Suasion*, Simon & Schuster, New York.
5. Holland, RW, Verplanken, B and Van Knippenberg, A (2002) On the nature of attitude–behavior relations: the strong guide, the weak follow, *European Journal of Social Psychology*, **32** (6), pp 869–76.
6. James, W (2015) *Principles of Psychology*, vol. 2, Andesite Press, London, p 463.
7. https://www.1843magazine.com/content/features/wanting-versus-liking (last accessed 25 August 2016).
8. Berns, G and Moore, SE (2010) A neural predictor of cultural popularity, available at SSRN 1742971.
9. Quoted in: http://news.bbc.co.uk/1/hi/health/8274773.stm (last accessed 25 August 2016).
10. https://nudges.wordpress.com/2009/03/23/nudge-for-sweet-teeth/ (last accessed 25 August 2016).
11. Halpern, D (2015) *Inside the Nudge Unit: How small changes can make a big difference*, Random House, London

Designing for screens

08

Figure 8.1 Number of hours, on average, people spend daily with each screen type

One of the more interesting image tricks to have appeared online in recent years is the split-depth gif. These are simple animated images, using three panes, that create a surprising three-dimensional effect on ordinary screens. For example, in one of the more effective split-depth gifs, a cartoon dog runs from left to right, moving from one pane to the next, then appears to run outside the right pane, around the white frame, over some text under the image, before darting back into another pane.

They work by harnessing a couple of visual cues that we use to judge whether we are seeing something in two or three dimensions. First, they often feature things that are moving from out of focus (in the background) into focus (as they apparently move towards us). Of course, this is often seen in videos, and of itself would not be enough to create a 3D effect.

The second trick, however, is cleverer. The images are typically divided into three equal sections by white parallel horizontal lines. The lines create a sort of extra set of frames around the images. Our brains – used to expecting images to stay within a frame – are momentarily surprised by seeing an image penetrating its border, and a slight three-dimensional effect results. In the dog example above, the effect is heightened by the dog running over the text caption below the image. We typically view the text caption below an image or animation as not being part of it, so this further enhances the impression that the dog has burst out of the image. (See Figure 8.2 for an idea of what split-depth gifs looks like.)

Figure 8.2 An example of what a split-depth gif looks like

We have all learned that frames are a sign that we should look inside them. Whether it is a window frame, the frame of a painting or the frame around a screen, they act as a cue to look inside. However, we also learn that most frames are showing us something that either isn't real or that is a simulation. This means that whilst frames can grab our attention, they can also keep us feeling slightly removed from what we are seeing. Frames or other borders (sometimes called 'holding devices') can act to draw our attention in, but selectively having a design element penetrating out of the plane can also make it look more dynamic, as though it were bursting out from the artificial world of the frame into reality.

Frames can also be used to boost visual saliency. Frames can increase the amount of contrasts in an image. For example, a frame of one colour against a background of another colour creates a zone of contrasts around what you want people to look at. As we saw in Chapter 6, areas of high contrast are more visually salient, and hence more likely to be looked at.

Of course the most frequently seen type of frame in the world today is that around computer and mobile device screens. These now dominate our attention like never before. A quarter of our waking hours are now spent looking at electronic screens. We are used to looking at multiple screen formats: TV, tablet, mobile phones, laptops etc. Often our eyes are switching from one screen to another, such as when people are browsing the web on a mobile device whilst watching TV.

A 2014 study looked at the average length of time people across 30 different developed countries spend daily looking at different types of screens.[1] In 16 of them, including the UK, the United States and China, people spent over 6.5 hours a day looking at screens – a large proportion of their waking hours. This was made up of approximately:

- smartphone: between two and three hours;
- laptop/PC: around two hours;
- TV: between 1.5 and two hours;
- tablets: between 0.5 and two hours.

TV, previously the king of screens, has now been relegated to a more secondary role in our lives. Even whilst watching TV, many people are now also using another screen – such as a smartphone or tablet – simultaneously.

Most of our screen time is now spent on interactive devices, rather than the more passive activity of watching TV. So our relationship with screens is mainly one where we are in control, often making choices about what we want to look at (such as on the web) and clicking away quickly if we don't think it is what we want.

Often that control means we are quick to click away from anything that does not immediately capture our interest, or that one screen is not enough to fully engage us: we flick our eyes between two screens, such as the TV and a mobile device. If web users already have a short attention span, the problem is made worse by multiscreening. Using more than one screen at a time divides our limited attention. Psychologists call our attention span 'working memory', and it is limited to three to five chunks of information.[2] If we are shifting our attention between screens we are dividing our attention power. This places pressure on screen imagery to be as engaging and immersive as possible.

Designing for screens also has its own idiosyncrasies. How people view things on screens and interact via screens can be different to real-life interactions. We read differently on screens, we behave with less social inhibitions when behind a screen, and we pay different amounts of attention to things on screens depending on how they are displayed.

Reading is harder on screens

For decades, experts have been predicting the coming of the paperless office. As electronic screens increased in size and quality, and decreased in price, people would not want or need to print so many documents. Or so some expected. Yet whilst we use screens more than ever, paper has not gone away.

Even when people have a mobile tablet, many still like to print things and read from paper. Many people also prefer reading paper books to e-books, for example. Interestingly, this preference for paper may not just be nostalgia for the charm of the printed word: there is evidence that we absorb information more easily when reading from paper than from screens.[3]

The definition and quality of screen images has obviously improved, with many screens now offering higher definitions than ever before, and e-ink formats on e-book readers looking very close to print on paper. Yet paper still offers advantages over even high-definition screens.

Viewing and reading on screens can be more distracting than from print. First there is the obvious ever-present temptation on most devices to browse away to somewhere on the web, or to open a game or app. Reading from a book, in contrast, can be a more focused activity. It is more suited to deep reading: concentrating on following a series of points or arguments, and thinking carefully about them.

Equally, reading down a webpage or on a document on a computer requires us to scroll almost constantly. We have to control either a mouse or keys to scroll the page down, and then as it moves down the text is in motion and we have to then readjust our eyes. All this may seem easy and simple, but it eats up a certain amount of mental resources. Psychologist Erik Wästlund, who has studied reading behaviour on screens, believes that this intermittent interruption can harm our reading by interrupting the flow of information in our short-term memories.[4] As we read, in order to understand the flow of ideas we have to hold them in our short-term memory in order to connect them together. We have to understand a sentence, then connect it to the meaning of the previous sentence, and previous paragraphs, keeping track of the overall structure.

The reading process is more interrupted by scrolling down a screen than flicking a page of a book. Even with e-readers we miss the tactile feedback that we get from holding a book, and seeing where we are physically within the book. Progress bars on e-readers attempt to compensate for this, but they are not so intuitive. Screens are more suited to reading by skimming and scanning, rather than deep, considered reading.

Given that reading online is a little harder than from paper, and that as web users we are fickle and impatient, it is not surprising that a lot of text on a webpage gets completely ignored. Web analytics company Chartbeat have accumulated a lot of experience measuring how far people scroll down a news story or article online.[5] It is typical for around four in 10 people to leave the page almost straight away, whilst most will only scroll down to around 60 per cent of the article.

Interestingly they found that only reading part of an article is no barrier to sharing it: there was no relationship between the most read articles and those most tweeted. Many were sharing articles without fully reading them. (And, presumably, those who then click on the shared link don't read the full article either!)

This is not necessarily a reason for only ever writing short copy – merely to be aware of the heightened importance of grabbing attention and conveying information early on. Design and images can help make up for this weakened form of attention. It is important to convey as much information as intuitively and easily as possible. Remember that people reading on screens are often dividing their attention and are skim reading. The techniques described in Chapter 3 can be helpful here.

However, in some instances there can be a behavioural economics-style advantage in including lots of writing. The subtext is that a lot of effort has been made, therefore it builds trust. For this reason, eBay listings with lots of text do well, even if all the text is not strictly necessary.

Ways to increase readability of text

There are also patterns to how people react to different layouts of text. Research shows that people read faster when there are more words per line (around 100 characters per line is best), but they prefer text laid out with fewer characters per line (45 to 72).[6] When people are more motivated and you need to convey a lot of information to them, 100 characters per line may be best. When you need to encourage them to read your text, keeping it concise – between 45 and 72 characters per line is probably preferable.

Given that people scan text online it can help to polish the clarity of your writing. One tool that can help with this is the Flesch–Kincaid readability formula. This tests how easy your text is to read by taking into account the length of your sentences (the longer the sentence, the harder to read) and the number of syllables in your words (the more, the harder). You can find versions of this tool online, for example:

http://www.readabilityformulas.com/free-readability-formula-tests.php.

It may be obvious but, particularly for websites that will be viewed on mobile devices, choosing a font that looks as large as possible is a good idea. This does not necessarily just mean increasing the point size of the font you are using. Some fonts just naturally look larger than others, even in the same point size. Experiment with how different fonts look on mobile devices.

Hard to read = hard to do

Designers now have a seemingly endless variety of fonts at their disposal. A font can, obviously, be chosen for its aesthetic appeal or for the general feeling it conveys: for example, basic fonts to convey a feeling of seriousness; a highly stylized script to convey old-fashioned heritage; or a cartoonish font to convey playfulness. There is even a popular documentary dedicated solely to the influence of the font Helvetica (simply titled *Helvetica*).

However, the principle of processing fluency that we looked at in Chapter 3 applies as much to fonts as it does to images. Key text accompanying designs, such as titles or descriptions, can aid or harm the overall feeling of fluency of a design according to how easy the font is to read. One study gave participants a set of written instructions on how to perform a particular exercise routine. Half the participants saw the instructions in a very clear font (Arial) whilst the other half saw them in a harder to read font (a brush script).[7] They were then asked to rate how long they thought the exercise would take, how hard it would feel, and how likely they would be to do it.

Those who read the exercise instructions in the easier font reported shorter estimate times, thought it would be easier and were more likely to do it than those who read it in the more difficult font. People were non-consciously using a System 1-type shortcut, using their own feeling of ease or difficulty in reading the instructions as a proxy for how easy the exercise would be! This has obvious implications for designing not only instructions, but things like descriptions of products and services.

Interestingly, however, there is one surprising benefit to disfluent fonts: they make it more likely that we will learn the information. Research in labs and classrooms has shown that the simple change of making a font slightly harder to read leads to better retention of the material presented.[8] This is probably because we are forced to read harder fonts more carefully – we pay more attention to them. This more careful reading style means we process the information more deeply and thoroughly.

There is a myth, dating back to the 19th century, that words are harder to read in capital than lower-case letters. The theory was that lower-case letters have more shape variety and hence create a more unique overall shape to each word, which then helps us recognize it. Newer evidence shows that this is not true.[9] People do read words in all capitals more slowly, but simply because we are less used to this format.

Video ads and memory

Have you ever walked into a room to fetch something and discovered that as you walk through the door you forget what you are looking for? This might not just be due to a bad memory, but may be an integral feature of the way we process information. As we move through our day we tend to group events together into chunks. For example, you may be sitting watching TV in one room and you hear your phone ring in another room. You go and answer the phone and have a conversation with a friend. Then you go into the kitchen and make a cup of coffee. Your brain will recall these as three separate events. Psychologists call this 'event segmentation'. It's a little like your brain dividing up your perceptions into movie-like scenes, with cuts in between. Your brain builds a little model of the scene so it can track what is going on. The model might include things like the physical layout of the room, the key objects in the room and the interrelationship between the characters who are interacting.

The moment when an existing scene is ending or a new one beginning is called an event boundary, and information at these boundaries is more likely to be remembered later. However, as we move across an event ▶

boundary, information from the earlier scene falls out of our immediate attention and becomes harder to recall. Our brain temporarily drops the old model of the scene in order to build a new one for the next scene.

Called the doorway effect, this helps explain why we often forget why we have moved into another room as we walk through the door.[10] However, it also happens when we are reading. The implications for designers are that when you have a narrative flow, such as in a video, or telling a story in writing, information that is at the beginning or end of a scene is more likely to go into long-term memory.[11]

In videos, information at the beginning of a scene, or approaching the end of a scene, is more likely to be remembered. But if someone has already perceived that the scene has ended, any information presented before the next scene could 'fall through the cracks' and be less remembered. This highlights a potential problem with the typical way that video ads are structured. The branding in ads is usually concentrated at the very end. However, if people feel as if the story of the scene has already finished, their brain may be distracted with processing the last scene, and waiting for the next. The branding could be less likely to be remembered later. An alternative approach that could offer a solution to this is called brand pulsing. Rather than concentrate the brand information purely at the end of the ad, brand pulsing involves putting the brand throughout the ad. This does not have to be too overt or heavy-handed, just little discreet appearances can be enough. It can even be effective to include elements that remind people of the brand through the ad, such as shapes or colours that are associated with it.

Similarly we remember scenes in videos according to how our brains cut them up. An event boundary in a video is not typically triggered by an editing cut, but by a change in the background or the scene itself.[12] This is related to two other quirks of memory: the Zeigarnik effect, and the peak-end rule.

The Zeigarnik effect is the fact that we are more likely to hold things that are incomplete in our memory than things that have finished. Incomplete tasks, for example, play on our minds until we have done them and they can be forgotten. Waiters are more likely to remember a table's order if they haven't yet paid. The Zeigarnik effect shows that leaving information incomplete, for the viewer to fill in the blanks, can help them to remember it more.

The peak-end rule is that when asked about a past experience, people's evaluation is most strongly biased by how they felt at the peak emotional moment of the experience and at its conclusion. When designing experiences it is worth thinking about how to build in a peak moment, and leave the person on a high at the end.

Disinhibition effect

Many of our daily tasks that used to involve face-to-face interaction are now conducted via screens. This brings with it a greater feeling of anonymity and creates something that psychologists call the *disinhibition effect*. This is the tendency for people online to feel less inhibited by social constraints than when interacting with others in person. For example, people are more honest in online questionnaires than when someone is asking the questions in person.[13]

Screens remove the feeling of self-consciousness or potential embarrassment that is ever-present when interacting with a live person. Of course, the more unsavoury side to this effect is the sometimes aggressive and anti-social way that people interact with others online. However, it can create other effects too. People can behave more honestly online, being more likely to give their honest opinions (for example when giving feedback reviews).

It also means that when buying things people consider a more diverse range of products from screens than in face-to-face retail situations. For example, a Swedish study showed that people were more likely to choose hard-to-pronounce alcohol products when buying online than when buying from behind a counter. (From a neuroscience perspective, giving a product a hard-to-pronounce name is probably a bad idea anyway! Not only are we less likely to talk about them, but names that are harder to pronounce feel more disfluent, and this could make them feel less intuitively 'good' to consumers.)

Research has also shown that people ordering pizzas online choose a wider variety of toppings, and order pizzas with more calories than if they are making their order in person. Ordering online removes the potential social stigma of being seen as unhealthy or overindulgent.[14] As well as the disinhibition effect, people may be more likely to feel comfortable making complex orders online where they can check their order easily, rather than when speaking to a person – where there is the possibility of them misunderstanding the order or making mistakes.

When needing to convey information that people may be too embarrassed to read in a public place, or ask for face-to-face, a website may be the ideal place. And with the access-anywhere nature of the web, delivering information to people at the moment they need it is now easy.

Mobile screens

With more than 2 billion people now owning a smartphone, the mobile web is the largest consumer marketplace in the history of mankind.

Mobile screens are obviously more versatile than laptops, desktops and TVs. The fact that they are almost always to hand means that they get looked at in more times and places than other screens. For example, when a person has time to kill, such as when they are waiting in a queue; or when they need 'just in time' information, such as needing to book a taxi or a restaurant table whilst out.

When people adopt shopping on smartphones they end up spending more with an online retailer and make more frequent purchases, possibly simply because they now have more opportunities to buy. Smartphones let people shop wherever they are, increasing convenience.[15]

Retail websites can take advantage of touchscreens. When users are able to touch a product, albeit on a screen, it feels closer to touching it in real life than simply clicking or dragging a cursor over it on a laptop or desktop computer.[16] Whilst the experience of touching items on a mobile screen such as a tablet is obviously not as rich as touching the real product – it doesn't give us a feeling of weight or texture – it nevertheless starts to make the experience more immersive and closer to real life. This is similar to the split-depth gifs mentioned at the beginning of this chapter.

Screen sizes also affect the ways that devices are able to persuade us. For example, larger smartphone screens are obviously better for watching videos than smaller smartphone screens. Research has compared users' responses to seeing web ads on large (5.3 inch) and smaller (3.7 inch) smartphones and found that people reacted differently.[17] Firstly people were more likely to trust video ads on the larger screen, and text ads on the smaller screen (text ads also worked well on the larger screen). Trust is an important factor in driving purchase intent, particularly if people don't know your company, or haven't bought from you before. However, the effect was not just on the amount of trust, but the type of trust. With the larger-screen phones, watching video ads was better at generating emotional trust. With the text ads on small screens the text ads were better at generating rational trust.

Emotional trust is the feeling of emotional security and positive feelings towards being a customer of a company. Rational trust creates greater confidence in a person's or an organization's competence and reliability.[18]

The researchers found that this was because of the different ways participants processed information on the two screen sizes. The larger screen induced a more System 1-type response, meaning that people made quick intuitive judgements. This was because watching a video on it was more immersive and triggered a greater multisensory feeling of involvement with the video, closer to the feeling of 'being there' than when seen on the smaller screen. They also speculate that, because people's brains were more occupied with processing the more complex visual and sound information, they were more likely to slip into the energy-saving System 1 mode of thought. In contrast, with text ads on the smaller screens, people were more likely to think rationally and critically – System 2 thinking – about the ads they were reading.

If you need to make a user feel good about doing business with you, and generate a general feeling of security, use video or rich imagery. Making a page personalized, using a person's name or other details, could help make it feel more immersive, as could allowing the person to input information through the device's microphone, adding in more sophisticated touch features on the page, or using the ability of smartphones to vibrate to add in moments of tactile feedback. If you need to reassure a web user that you are competent and reliable, use lots of text. Of course, a combination of both may boost both types of trust, but with limited consumer attention spans, particularly if they are using a mobile device on-the-go, you may need to focus on one or the other.

As more sophisticated smartphones become popular, it may be possible to induce more immersive 'being there' feelings to webpages and ads through stimulating people's sense of touch. This could include effects like better, more fluid interaction with products on touchscreens – such as being able to rotate 3D high-definition images.

As well as creating more immersive experiences on screens, it is also helpful to understand our inbuilt biases on where our eyes are drawn on screens.

Gamification

Computer and smartphone games use a variety of techniques to engage people in tasks that can often be quite difficult. We play them for their own sake (ie non-gambling games that have no potential financial reward) because they are fun and give us a sense of achievement. Gamification is the practice of taking game elements and introducing them into other areas, such as using emotionally engaging or fun animations and sound effects, or allowing users to compete by collecting points or rewards.

Gamification can be particularly useful if you need people to maintain a behaviour over time. It can be used to reward loyalty, or progress on a health or fitness regime, for example.

Central fixation bias

In the Middle Ages, soon after Gutenberg invented the printing press, the method for laying out printed pages was kept a trade secret. However, their methods were essentially to create a print area on the page that was in the same proportions as the page ratios, similar to the self-similarity effects we looked at in Chapter 3. This undoubtedly helped their pages to look more intuitively attractive.

Today, the web also has its own hidden patterns that affect whether a page feels intuitive to a viewer. When presented with arrays of images – such as on a webpage or a range of products on supermarket shelves – people are biased towards looking at and choosing the middle and avoiding the edges. There is also a bias towards the top left.[19] Psychologists call this the central fixation bias.

There may be different factors behind the central fixation bias. For example, on screens or pack designs people may expect to find more information in the middle of the design, as this is just what they have learned from experience with designs (see the section on 'prototypicality' in Chapter 4). We know that people's eyes will fixate quickly on the areas where they expect to find the most meaningful information (as opposed, for example, to areas that merely have the most detail). They are also more likely to return to fixate on these meaningful areas rather than fixate on new ones.[20]

Equally, in the real world where we can move around to view things, we may become accustomed to seeing the things we are most interested in in the centre of our field of vision, simply because we have positioned ourselves to best see them. There may also be an evolutionary reason behind the avoidance of the edges: people may have a non-conscious evolved fear of being on the edge of a crowd – where predators could more easily attack them.

The central bias seems to be a general phenomenon, no matter what people are looking at. For example, in one study of radiologists looking for nodules in CT scan images of lungs, more than eight out of 10 of them failed to notice an image of a gorilla that had been inserted into the top right of the scans.[21] This was despite the fact that the gorilla image was nearly 50 times the size of an average nodule. If even trained image observers like radiologists miss things in displays, the average web user certainly will.

Whether people are biased towards the centre or the top left may depend on whether there is an obvious centre to the array. For example, in a three-by-three array of images there is one image in the centre, and there is evidence that most people will look at that image more than the others. However, in a four-by-four array there is no one image in the middle. Instead there are four images in the middle, and it is the top left of those four images that gets the most views (see Figure 8.3). Similarly when there is a two-by-two array, again with no centre image, it is the top-left image that gets the most attention.

Figure 8.3 When items are in a grid arrangement, if there is a centre position (such as in the grid on the left) we tend to look there first. If there isn't (such as in the grid on the right) then we tend to look first at the top-left position

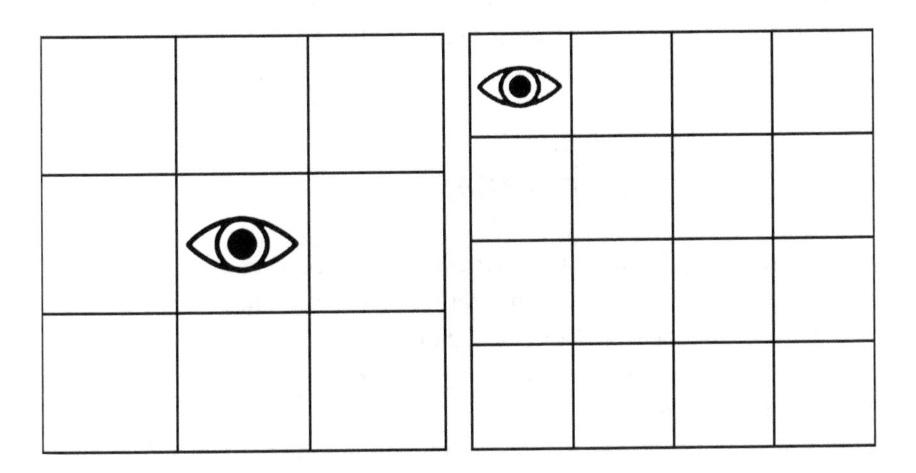

These visual biases can also translate into how people make decisions. Just as we saw with people being more likely to pick items that are more visually salient, items presented centrally are more conspicuous and more likely to be considered.

These biases are important on webpages as images are often presented in arrays, sometimes as a menu format. The image that you want people to look at first, or gain most attention, should be placed in the top-left position when there is no centre position, or the centre when there is one. The effect is similar to that of visual saliency: biasing where people will look. Combining the two strategies – making an image more salient, and placing it into the centre or top left of an array (as appropriate) – could give it an extra boost of attention.

Screen aspect ratios

Until around 2003, most computer screens were the same aspect ratio as old TVs: 4 to 3 – almost square. Then for a brief few years the industry moved to an almost golden ratio of 16 to 10, likely making screens more attractive and intuitive to view. By around 2010 they changed again, but less dramatically, moving to a 16 to 9 ratio, which is better for watching movies. However, screen sizes are measured by their diagonal length from one top corner to the opposite bottom corner. So a screen size of 28 inches with an aspect ratio of 4 to 3 actually has a larger viewing area (250 square inches) than a 28-inch 16 to 9 screen (226 square inches).[22]

Horizontal viewing bias

People also have a horizontal viewing bias when viewing displays. We find it easier to scan items from left to right (or vice versa) than up and down.

When viewing images of all kinds, people tend to look first to the left of the image, and then move to the right. This means – as we saw in the phenomenon of pseudoneglect in Chapter 3 – that the left sides of images get looked at more, and information on the left is exaggerated more in our minds than that to the right. It also means that we are more likely to look left to right than up and down. This obviously has implications for things like on-screen menu design. A horizontal menu is probably usually better than a vertical one.

Evidence also shows that we make more sideways fixation patterns than vertical ones.[23] From an evolutionary perspective this makes sense. We have two eyes, positioned horizontally next to each other. Our way of seeing things is more optimized for a horizontal gaze than a vertical one. Our ancestors would also have needed to look left and right more than up and down. For example, scanning the horizon would have typically been done more often than looking from the ground to the sky. Designs that require we look side-to-side probably feel more fluent to us than those that make us look up and down. This may also help explain the preference for horizontal rectangles. The horizontal rectangle presents information in the way we naturally prefer to see it.[24]

It is also worth noting that this may give a disadvantage to people seeing things on a smartphone. The rectangle screens of smartphones are typically orientated vertically (portrait) rather than horizontally (landscape), due to this being a more natural way to hold one in your hand.

If you need to present information on the right side of the screen, or more vertically than horizontally, and get it noticed, there is a technique for counteracting the left-side and horizontal biases. Placing the images or text inside a box or frame will help draw in people's gaze.[25]

CASE STUDY Netflix

Understanding how people react to images on screens is critical to the success of Netflix, the online streaming service for movies and TV shows. What makes people decide to watch a movie or TV show? Netflix has spent several years researching this. Being an online, interactive platform they are in an excellent position to track what people look at, and what they then decide to watch.[26] The first interesting finding is that 82 per cent of users' browsing was directed at images rather than text descriptions, underlining the way that people often rely on images more than words when browsing and making decisions. The quality of the images was also important in convincing users to watch something.

Netflix's research revealed several big findings on how their users react to images:

1. **Images of three people or fewer**
   Whilst the appeal of many films and TV shows is a large complex cast, when browsing users prefer preview images with only three or fewer people. This is similar to the principle of *subitizing* that we looked at in Chapter 3: our ability to process a small number of items automatically; so when there are more than four or five items – such as people – in an image, we have to process them one by one.

2. **Complex facial emotions**
   We have seen in Chapter 5 how important images of faces are in engaging people. What Netflix found was that complex and subtle expressions on faces were particularly effective at drawing users' interest. As we are so adept at decoding emotions and intentions from people's faces, it is possible to convey many different ideas or emotions that will be featured in a show, just from a face.

3 **Showing the villain**

Perhaps surprisingly, showing the villain was more effective than showing the hero. This was true whether it was a children's show/movie, or a regular action movie. Showing the villain probably gives viewers a better feel for the conflict that is going to be shown.

Whilst screens lack the advantages that paper has for deep reading, they present many opportunities for presenting information. They lack the physical, tactile intuitive feedback of paper and books, but they can introduce more immersiveness through rich imagery, videos, touch and vibration effects (on mobile). By using some of the techniques in Chapter 3 for making designs more intuitive, you can also overcome the divided attention spans of viewers.

With most designs now being viewed online, it is now also easier than ever to test them. A/B testing online is probably the easiest way to test images. On any webpage, two or more versions of an image can be used for different users, and you can measure any differences in click-throughs or times spent on the pages with those different images.

Summary

- Reading text is harder online than on paper, and we read with less depth on screens.
- When viewing narratives on a video, or reading a narrative, people mentally process it in terms of discrete events or scenes. Information is more likely to be remembered if it happens at the beginning or end of a scene, but beware including important information in between scenes, as it is less likely to be stored in memory.
- In general, use easy-to-read, clear fonts. However, if you need to present information that you want people to concentrate on and remember later, use a more stylized or slightly harder-to-read font.
- People can be less inhibited by social judgements online – a phenomenon called the disinhibition effect. This can result in people buying things that they would be embarrassed to buy in person in a shop, or considering wider ranges of products (such as those with hard-to-pronounce names).
- When images are arranged in an array or grid, people have a bias towards viewing the central image, or if there is not an absolute centre, the top-left image.

- People also have a bias towards looking at the left before the right on a screen, and find it easier looking sideways than up and down. These effects can be overcome by putting text or images into a box or frame.

Notes

1. http://bgr.com/2014/05/29/smartphone-computer-usage-study-chart/ (last accessed 25 August 2016).
2. Cowan, N (2010) The magical mystery four: how is working memory capacity limited, and why?, *Current Directions in Psychological Science*, **19** (1), pp 51–57.
3. Mangen, A, Walgermo, BR and Brønnick, K (2013) Reading linear texts on paper versus computer screen: effects on reading comprehension, *International Journal of Educational Research*, **58**, pp 61–68. See also Dillon, A (1992) Reading from paper versus screens: a critical review of the empirical literature, *Ergonomics*, **35** (10), pp 1297–326; Gould, JD, Alfaro, L, Barnes, V, Finn, R, Grischkowsky, N and Minuto, A (1987) Reading is slower from CRT displays than from paper: attempts to isolate a single-variable explanation, *Human Factors: The journal of the human factors and ergonomics society*, **29** (3), pp 269–99.
4. http://www.wired.com/2014/05/reading-on-screen-versus-paper/ (last accessed 25 August 2016).
5. http://www.slate.com/articles/technology/technology/2013/06/how_people_read_online_why_you_won_t_finish_this_article.html (last accessed 25 August 2016).
6. Dyson, MC (2004) How physical text layout affects reading from screen, *Behaviour & Information Technology*, **23** (6), pp 377–93.
7. Song, H and Schwarz, N (2008) If it's hard to read, it's hard to do: processing fluency affects effort prediction and motivation, *Psychological Science*, **19** (10), pp 986–88.
8. Diemand-Yauman, C, Oppenheimer, DM and Vaughan, EB (2011) Fortune favors the bold (and the italicized): effects of disfluency on educational outcomes, *Cognition*, **118** (1), pp 111–15.
9. Rayner, K (1998) Eye movements in reading and information processing: 20 years of research, *Psychological Bulletin*, **124** (3), p 372.
10. http://www.scientificamerican.com/article/why-walking-through-doorway-makes-you-forget/ (last accessed 25 August 2016).
11. Swallow, KM, Zacks, JM and Abrams, RA (2009) Event boundaries in perception affect memory encoding and updating, *Journal of Experimental Psychology: General*, **138** (2), p 236.

12 Schwan, S, Garsoffky, B and Hesse, FW (2000) Do film cuts facilitate the perceptual and cognitive organization of activity sequences?, *Memory & Cognition*, **28** (2), pp 214–23.

13 Lind, LH, Schober, MF, Conrad, FG and Reichert, H (2013) Why do survey respondents disclose more when computers ask the questions?, *Public Opinion Quarterly*, 77 (4), pp 888–935.

14 Goldfarb, A, McDevitt, RC, Samila, S and Silverman, BS (2015) The effect of social interaction on economic transactions: evidence from changes in two retail formats, *Management Science*, **61** (12), pp 2963–81.

15 Wang, RJH, Malthouse, EC and Krishnamurthi, L (2015) On the go: how mobile shopping affects customer purchase behavior, *Journal of Retailing*, **91** (2), pp 217–34.

16 Brasel, SA and Gips, J (2014) Tablets, touchscreens, and touchpads: how varying touch interfaces trigger psychological ownership and endowment, *Journal of Consumer Psychology*, **24** (2), pp 226–33.

17 Kim, KJ and Sundar, SS (2016) Mobile persuasion: can screen size and presentation mode make a difference to trust?, *Human Communication Research*, **42** (1), pp 45–70.

18 Johnson, D and Grayson, K (2005) Cognitive and affective trust in service relationships, *Journal of Business Research*, **58** (4), pp 500–07 (in the research literature, emotional and rational trust are referred to as affective and cognitive trust – I have just changed the terms for the sake of simplicity and clarity).

19 Falk, R, Falk, R and Ayton, P (2009) Subjective patterns of randomness and choice: some consequences of collective responses, *Journal of Experimental Psychology: Human perception and performance*, **35** (1), p 203.

20 Kapoula, Z, Daunys, G, Herbez, O and Yang, Q (2009) Effect of title on eye-movement exploration of cubist paintings by Fernand Léger, *Perception*, 38 (4), pp 479–91.

21 Drew, T, Võ, MLH and Wolfe, JM (2013) The invisible gorilla strikes again: sustained inattentional blindness in expert observers, *Psychological Science*, 24 (9), pp 1848–53.

22 Barrow, JD (2014) *100 Essential Things You Didn't Know You Didn't Know about Math and the Arts*, WW Norton & Company, London.

23 Ossandon, JP, Onat, S and Koenig, P (2014) Spatial biases in viewing behavior, *Journal of Vision*, **14** (2), p 20.

24 http://www.independent.co.uk/news/science/why-some-shapes-are-more-pleasing-to-the-eye-than-others-1847122.html (last accessed 25 August 2016).

25 http://www.sciencedirect.com/science/article/pii/S0042698912003914 (last accessed 25 August 2016).

26 http://www.fastcompany.com/3059450/netflix-knows-which-pictures-youll-click-on-and-why (last accessed 25 August 2016).

Viral designs 09

Figure 9.1 Images can spread online in the same way as a virus spreads in nature

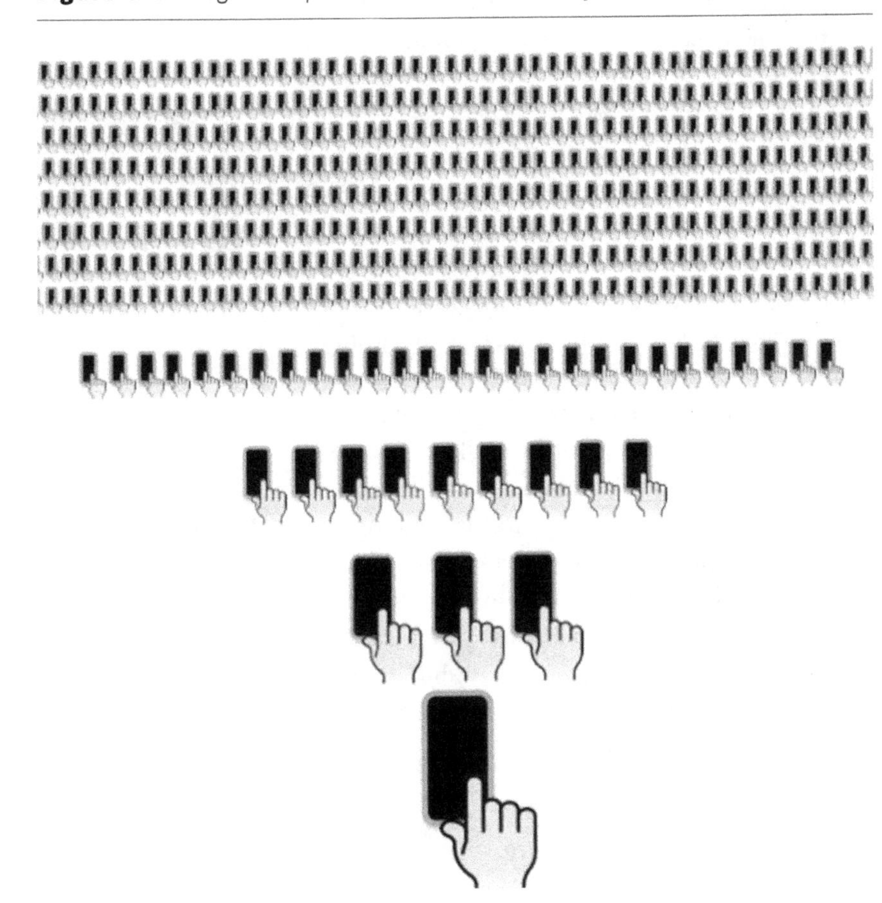

During the First World War a curious image began appearing on walls in air force camps and on the sides of railway carriages. Australian servicemen had been drawing a graffiti-like doodle declaring that 'Foo was here' – a spontaneous example of what we would now call viral advertising, except it wasn't advertising anything other than the fact that a serviceman had been in that location. These viral images were copied even more during the Second World War, with US servicemen writing 'Kilroy was here', and the British version declaring 'Chad was here'. The writing was accompanied by a doodle of a bald man peeking over a wall, with his long nose overlapping it.

Interestingly the nose of Foo/Kilroy/Chad is a clothoid curve. Clothoid curves are like a stretched 'U'. They can be seen in designs as varied as rollercoasters and off-ramps on major roads that allow drivers to turn back in the opposite direction. They are also very common when people are making casual doodles. The reason why this type of curve is used in these three examples is because it is the easiest curve shape for a moving object – be it a car, rollercoaster or a hand directing a pencil – to make whilst moving at a constant speed.[1] Clothoid curves are just easier to draw; they have processing fluency for the person drawing them.

However, the more important point about Foo is that he is an early example of a viral image. The doodle spread spontaneously, without any single person or central organization directing it. Also, it has a couple of things in common with many viral images we see on the web today: it is based on humour and on a face. However, whilst in previous chapters we have examined things that make images attractive or aesthetically pleasing, we will see that this can actually harm an image when it comes to making it viral. Foo/Kilroy/Chad is easy to draw, but it's not especially easy on the eye or beautiful. Nevertheless this doesn't matter when it comes to making it propagate virally.

Memes

In general, images and ideas that go viral have become known as memes. The concept of a 'meme' was introduced in 1976 by the biologist Richard Dawkins in his book *The Selfish Gene*.[2] Whereas a gene is a batch of biological information that copies itself via living organisms reproducing, a meme is, analogously, an idea that evolves and propagates through human culture. To extend the analogy: in the same way that the challenges of the environment put pressure on a species to evolve, memes evolve through being useful to us culturally. This usefulness can simply be that they amuse us, help us to communicate or bond socially. The notion of memes has even inspired its own scholarly field – mimetics – that seeks to understand how memes spread, using evolutionary and other scientific models.

The essence of a meme is that it gets adapted and copied. Not all ideas become memes. An ancient myth that has survived for millennia would be a good example of a meme. Understanding how genes spread requires studying biology; understanding how memes spread requires understanding the mind. Visual memes are therefore an important subject for neuro design.

Examining memes and what makes them work can provide insights into making popular designs that propagate under their own power. People want their images to go viral. Viral marketing can be supercharged online, as the most contagious images can grow and propagate like a real virus, with none of the costs associated with advertising (previously the only real way to reach a mass viewership).

Futurist Alvin Toffler in his 1980 book *The Third Wave* coined the term 'prosumer' to describe the blurring of the roles of consumers and producers.[3] Decades on, we can see how the web has provided a medium for this. Hundreds of millions of people globally now upload their own visual content, some completely original, others mixes or mash-ups of pre-existing images or videos. Nearly 2 billion images are now uploaded on to the web every single day.[4] Visual memes are a prosumer phenomenon. They may originate as pop-culture images (eg photos of celebrities) but they are then adapted, added to or simplified. Hundreds of hours of video are uploaded every minute to YouTube alone, much of which is user generated or user mash-ups.

The viral advantage of mash-up imagery is that it comes with built-in recognition: using images from pop culture means that others will recognize them. They have familiarity appeal. They also can propagate more effectively if they are introduced into a pre-existing network of fans or interested people. For example, a funny quote added to an image from *Star Trek* gets shared amongst social media blogs, groups and Twitter fans of that fictional world. Or they propagate because they tap into something with widespread appeal, such as the cuteness of cats or babies.

So successful viral online imagery has two parts: social and visual. The social influences are things like how many followers the person or organization who posts the image/video has, whether they are liked/trusted and so on. Then there is the effect of the image itself: whether it appeals to people. As there is a mixture of these two effects this can make viral visuals challenging to study in a controlled way. For example, an image may become successfully viral because it is inherently designed to trigger sharing with others, or it may become viral because it gets lucky and people with large social media followings decide to share it. Equally an image might be shared if it relates to something currently in the news or if it has a celebrity in it. Yet this is not necessarily that useful to a designer. The type of images we are interested in here are those that become viral due to the design itself, not its social content. We need to look more to the real world to understand what makes a successfully shared image.

Internet memes

What do Barack Obama, astrophysicist Neil deGrasse Tyson, musician Freddie Mercury, and the actors Jackie Chan, Patrick Stewart and Nicolas Cage have in common? The answer is that cartoon versions of their faces have all become popular memes online. Each caricature perfectly captures a feeling or social moment that we can all relate to. These images then get posted on discussion boards, comments sections on websites and in social media posts as a shorthand way to respond to a topic under discussion. They are a little like emoticons: a widely used way to express feelings graphically. These particular celebrities have probably become popular as memes due to their high recognizability amongst the young men who predominate on anarchic online discussion forums such as Reddit and 4Chan, where many of these memes first originate and spread.

Figure 9.2 Viral internet memes

These memes typically portray a feeling that either does not have its own word, or cannot be fully conveyed with just a word. They are similar to a new linguistic form that has become popular with the rise of social media: 'That moment when…' This is a new type of sentence construction that has arisen spontaneously to help people communicate feelings that are often specific to a particular occasion or event, or that by just describing the occasion of the event others will smile and nod along in recognition. For example: 'That moment when you realize you've forgotten your keys', or 'That moment when you can't stop laughing whilst telling a joke, but when you finish none of your friends laugh.' Linguists call them a subordinate clause – they are not complete sentences so they require you to fill in the blank, like a simple puzzle. In other words they leave unspoken the missing 'Remember how it feels when…' or 'You know how you feel when…'

As we have seen, the brain loves simple puzzles, they help evoke a blip of pleasure as we resolve them. Also puzzles that invite us to fill in the blank are using the Zeigarnik effect: incomplete things create a tension that makes them linger in our minds until we have resolved them. It is hard to read a subordinate clause without automatically filling in the missing part. Subordinate clauses also have the inbuilt assumption that the feeling they are describing is universal. By writing 'That moment when…' but not explicitly defining what the feeling is, you are trusting that others know the feeling too. It's a pre-eminently social form of writing.

Whereas memes were originally defined more generally as ideas that propagate and evolve, the term has now taken on this more specific meaning: humorous cartoons or photos that typically capture an easily recognizable feeling. Sometimes more general trends such as particular genres of YouTube videos or social media hashtags get described as memes, but the main meme is the humorous feeling-based image.

Users often adapt these memes to a current subject by adding in text. For example, they might be commenting on their reaction to a current piece of celebrity news, or just telling a short-form story of something interesting or funny they have done. This helps the meme image to spread by making it versatile enough to fit different situations; it is adaptable. Like real biological genes, they can mutate, and these mutations can help them to spread even more.

The most viral images on the web

As at the time of writing, the top 20 most viral images of all time online most prominently feature faces, either human (13 of the images) or animal (seven images); 12 could be considered humorous; and only three feature celebrities or famous people. Animals appearing eerily human feature in six of the images (for example, a cat with a relatable apparently grumpy facial expression). Animals 'photo-bombing' – jumping into the frame of a photo – appear in two of the images, such as a manta-ray popping up behind a group of bathing women and apparently embracing them, and a squirrel popping up in a photo of a couple in the wilderness, and looking directly into the camera. Like animals with relatable facial expressions, photo-bombing animals create both humour and the unexpected familiarity of an activity that we usually associate with humans. Whilst marketers often assume that it is beauty and physical attractiveness that are the most powerful drivers of popular images, only two of the top 20 images are explicitly focused on the physical attractiveness of the main subject.[5]

Memes and neuro design

Internet memes may be the ultimate form of neuro designs. They use a number of neuro design principles that we have covered in earlier chapters:

- **Peak shift effect** (see Chapter 2)
  Many memes are like caricatures. For example, the ones based on real people are, despite often being very simple line drawings, instantly recognizable. Whilst they typically lack the exaggerated features of standard caricatures, they capture the essence – or *rasa* – of the person, including a very typical facial expression or posture that the person has.
- **Minimalism** (see Chapter 3)
  Memes are very simple and easily understood. They often evolve into becoming simpler over time. For example, photos get turned into line-drawing versions. Whilst turning a photo into a line drawing helps enhance its peak shift effect, it also makes it more minimalistic. This minimalism gives the image neuro-appeal but also gives it a technical advantage: it works on a wider array of screen formats, even at lower resolutions.

- **Unexpected familiarity** (see Chapter 3)
  Animal memes tap into the power of unexpected familiarity. One popular meme on Reddit features a photo of a seal leaning its head back with its face frozen in apparent panicked embarrassment – the sort of feeling one has when realizing you've done something wrong or committed a faux pas and are anxious in case others discover it. This sort of worried anticipation is something we can all relate to, but we don't expect to see it on the face of an animal, therefore it has the appeal of unexpected familiarity.

  Equally, the feelings that these memes express often don't have a name. In other words, in order to describe the feeling you would have to construct a whole sentence, whereas just showing an image that captures the feeling in a quick-to-understand way feels far more fluent.

- **Prototypicality** (see Chapter 4)
  Meme images are often the raw, concentrated version of a feeling. In real life most of us will often feel something either without expressing it on our face or with a minimal version of the full facial expression. Also, we often have mixed feelings, so our facial expressions are not pure versions of that feeling. Memes are pure expressions of typical feelings we all have.

- **Emotional appeal** (see Chapter 5)
  Memes are almost all based on feelings. In particular, memes seem to make particular use of social emotions, such as embarrassment, incredulity or admiration. Social emotions are inherently shareable. Likewise humour has inbuilt shareability: if we hear, read or think of something funny we naturally feel the urge to share it. By doing so we bond with others and win social points.

- **Visual saliency** (see Chapter 6)
  When researchers analysed the popularity of over 2 million images on the site Flickr, they discovered that even the colour of images had an effect.[6] In general, images with more blues and greens were less popular than those with more reds. The researchers speculate that this is because more striking colours (such as red) will be more likely to grab attention – in other words, have higher visual saliency. As people scroll, flick and browse social networks it is not surprising that images that have high visual saliency are likely to have an advantage.

- **Social proof** (see Chapter 7)
  Being part of the 'in crowd' was probably a vitally important drive for our ancestors. In order to survive and have children we needed to bond with cooperative groups, family and tribe. Becoming an outcast could

easily have meant death. Sharing memes is a form of bonding. It communicates that you are part of the in-group because you are sharing the joke. Others will typically be familiar with the feeling your meme is expressing, probably the meme itself – and if it features a celebrity, with that person too.

The lesson to be drawn from these types of memes is the viral appeal of images that portray emotions and align with neuro design principles. There is probably a lot of untapped potential in finding new 'moments when...' or social emotions that can be illustrated in a minimal, recognizable way with simple faces and body positions.

Mimetic desire

As well as simply spreading ideas, desire or wanting can propagate. Some neuroscientists have called this mimetic desire. In one study participants lay in an fMRI scanner whilst they watched a series of videos.[7] The videos showed two identical objects – such as toys, clothes, tools or food items – the only difference being that the two objects were of different colours, and in the videos a person was shown selecting and picking up one of the pair of items. Afterwards participants were asked to rate how much they liked each object. The results showed that if they had seen someone select an object, no matter what the category of object, they were more likely to say they liked it.

The brain-scan data showed that two brain systems in particular became active at the same time when people viewed the videos. First, something called the mirror neuron system. This involves parts of the brain that not only become active when you prepare to make a movement (such as picking up something) but also when you see someone else perform that movement. In other words, just seeing someone else perform an action – such as picking up an item – can cause your brain to mirror that activity. The second area that became active was the system that anticipates rewards. The neuroscientists see the triggering of these two areas simultaneously as the marker of mimetic desire. Interestingly, just like the mere exposure effect we saw in Chapter 3, merely seeing someone pick up something was enough to trigger mimetic desire. It is also similar to the endowment effect that we saw in Chapter 7. The endowment effect is when we value something more once we own it, but mimetic desire is to want something more once we see others own it. It's a little like watching a small child play with other children: when they see another child pick up a toy, often they then want that toy too!

Emotions and viral content

Emotions play an important role in making content viral online. People are more likely to share links if they felt emotionally moved in some way. For example, web company Buzzsumo analysed the popularity of 100 million online articles and found some interesting trends.[8] First, as we would expect, the presence of an image increased the likelihood that the article would be shared (articles with images were more than twice as likely to be shared). They also analysed the emotional content of the posts and found that particular emotions were more likely to lead to shares. These were awe, amusement/laughter and joy, with anger and sadness being less shared.

Awe is an interesting feeling from a neuro design perspective. It is often overlooked by marketers but images that evoke awe can be very popular online. Awe is typically evoked by a combination of both beauty and large scale. For example, natural scenes that are vast and attractive, such as beautiful photos of the Grand Canyon. There is something about the feeling of awe that makes us want to share it with others.

They also found that infographics were highly shared. This is, again, predictable from a neuro design perspective. As I described in Chapter 3, infographics can be a good way to evoke the appeal of 'simpler than expected'. They are novelty and information rich, whilst (if they are well designed) easy and intuitive to ingest.

Another study examined the virality of around 7,000 articles from the *New York Times*, taking into account the emotional content of the articles.[9] In general, they concluded, emotional content was more viral than non-emotional, and positive emotions beat negative ones. Again, the feeling of awe was found to be highly shareable. Equally, positive but low-arousal feelings, like contentment or relaxation, are not enough to induce virality. It needs a more high-arousal emotion to move people enough to share the content.

Exceptions to the negative-emotions finding were anger and anxiety-inducing articles. If a news item makes someone angry, they may want to share it to rally opposition amongst their friends or those who share similar political leanings, for example. If something induces anxiety, sharing it with others may help the person to feel less anxious. However, articles that evoked sadness were unlikely to become viral.

Archetypal images

Swiss psychoanalyst Carl Gustav Jung believed that certain images resonate universally, deeply and non-consciously in us because they tap into our instincts. He called these images archetypes. He described them as 'primordial images that reflect basic patterns that are common to us all, and which have existed universally since the dawn of time'.[10] These images are not the instincts themselves, but express them, and hence resonate deeply with us. The exact form of the images can change and evolve over time, but they are still tapping or expressing the same underlying archetype. As they are universal, and built into our minds, if an image evokes an archetype it is both quickly and easily understood by viewers. Archetypes can be personality or character types, such as the magician, the trickster, the hero or the care giver. They can also be non-human things such as the monster or the forest. Archetypes are not really a neuroscience concept (eg we don't know how these instinctive forms could be inherited and stored in the brain, let alone triggered by images) but they provide an interesting model for speculating on the success and prevalence of certain types of images.

Can a computer predict if an image will go viral?

If the content of an image (and not merely how topical it is, or how lucky it gets in being passed on by an internet celebrity) can influence how viral it becomes, then it should be possible to train a computer to predict the viral potential of an image. This is exactly what one group of researchers have done. They put together a database of around 10,000 images that had been posted on Reddit, and for each image collected data on how popular it was (ie how many 'upvotes' viewers had given it, and how often it had been shared).[11] This allowed them to create collections of images that had been highly viral, those that hadn't been viral at all, and those that were somewhere in-between. They then showed a selection of different types of image to people and asked them to predict how viral they thought the image was. They also got people to rate images against 52 different attributes, such as types of emotions that the image evoked. They discovered that there are clear patterns of image content in viral images. The top 15 characteristics are:

1. Synthetically generated (eg a Photoshopped image)
2. Cartoonish
3. Funny
4. Animals
5. Explicit
6. Dynamic
7. Man-made
8. Cute
9. Sexual
10. Male
11. Weird
12. Alarmed
13. Scary
14. Old
15. Suspicious

Image content that was least likely to go viral included:

1. Relaxed
2. Open spaces
3. Beautiful
4. Calm
5. Tired
6. Serene
7. Objects
8. Sleepy
9. Depressed
10. Positive atmosphere
11. Centred
12. Pleased
13. Frustrated
14. Grouped
15. Colourful

Interestingly, symmetrical images were also unlikely to be viral – an example of when an image might be beautiful or aesthetically pleasing but does not evoke the desire to share. The list of non-viral image content shows that images that are beautiful, or convey low-intensity pleasant emotions (such as calmness or serenity), may be nice but are not shareable. The success of Photoshopped or synthetic images shows the value of mashed-up imagery. When images can be adapted by others they can be made more topical (eg adding text that relates the image to the topic under discussion or something in the news), or funny (eg face-swap images, such as a photo of an adult and a baby with their faces swapped). The success of this type of image could also be because they evoke unexpected familiarity (see Chapter 3). In other words they show something familiar, but in an unexpected way. Images that had more than one of the successful viral characteristics (eg were Photoshopped and included an animal) were particularly likely to be viral.

However, what if a designer needs to make a beautiful or low-emotional-intensity image viral? In other words, can an inherently non-viral image be made viral? The answer is yes. The researchers found that even if an image contained a non-viral subject, it could be made viral by adding in one or more of the most viral elements.

Of course, some of the types of content that made for viral images on Reddit might differ for other social networks or contexts. Reddit is heavily skewed towards a demographic of young males, and the fact that people can post anonymously makes more edgy or anarchic content more likely to be shared. Nevertheless, the findings broadly match what other research shows. Humour, cuteness, simplicity (eg cartoons) and high-intensity emotions (like alarm) are more viral, whilst beautiful and serene images are not.

The researchers also discovered that people could predict the virality of an image above chance rates (about 65 per cent) even if they hadn't seen it before. Their accuracy improved when they were shown pairs of images, one of which had been viral and one that hadn't, and were asked to guess the viral one (their accuracy here was around 70 per cent). However, they were less accurate when the pairs included one image that was in-between high and low virality (around 60 per cent). They then trained a computer to analyse images on the results from the 52 attributes, and the computers could predict a viral image better than a human could (65 per cent accuracy!). Of course, the computer had access to the hidden knowledge of what makes an image viral, but it is still impressive.

The main viral image sites

Social networks are the main breeding ground for viral imagery. Each of the main networks has its own characteristics, and different types of content can be more suited to different networks:

- **Facebook** (over 1.5 billion active users)
  Facebook is the largest social network. Images on Facebook can get both good engagement (ie people commenting and discussing them) and shareability (users reposting the image on their own Facebook wall). Viral marketing on Facebook is possible via a number of different visual media: images, animated gifs, videos or simple games. Facebook is primarily used by people keeping in touch with what their friends and family are doing, therefore they are primed to consume and share content that has a personal touch. Inspirational, fun or novel content that tells a story about a real person or people is a good fit for Facebook. Equally, as people now regularly check into their Facebook account to browse or upload photos throughout the day, wherever they are, more location-based viral marketing can work too. For example, if you have a business with a physical presence, can you create something novel or fun that people will want to take a photo of and upload to Facebook? One example of this is the growth of bars, cafes and restaurants writing funny things on their chalkboards outside. People may take a photo of this and upload it to Facebook, getting that establishment noticed.
- **Instagram** (over 400 million active users)
  Instagram is even more image-focused than Facebook. In particular it is used for showcasing high-quality photos. Its inbuilt range of filters has meant from the start that users try to improve or enhance their images before posting them. As it is more open than Facebook (images are shared with anyone more readily, not just closed networks of friends and family) there is more chance of people discovering images of things they are interested in, be they products, places, events or activities. This highlights the importance of tagging images on Instagram. Research shows that faces are a powerful driver of engagement on Instagram.[12] Photos with faces are about one-third more likely to receive likes and comments.
- **Twitter** (over 300 million active users)
  As Twitter only allows short posts of 140 characters, images can help squeeze in a lot more information. Twitter is a real-time stream of consciousness. Posts instantly appear in people's feeds but then quickly drop down as they are replaced with newer ones. Therefore posts work

most powerfully in the present moment. This enhances the importance of posting images that relate to current events and live topics of discussion, such as sporting or news events. Equally, asking for people to retweet your post (eg 'Please retweet' or 'Please RT') helps it to get shared, as does including a link (pointing to the value of sharing more information than just the narrow channel that Twitter allows within its 140 characters).[13]

- **Pinterest** (over 100 million active users)
  Pinterest was the fastest-growing social network to reach over 100 million users.[14] It is a vehicle for visual discovery. Rather than Twitter's real-time sharing of activities and opinions, or Facebook's community of friends and family, Pinterest users are hungry for discovering cute, useful and desirable things (as opposed to other social networks where photos of faces and people are more prevalent). Browsing on Pinterest is a little bit like window or catalogue shopping. The Pinterest community is about 80 per cent female. Pinterest is also more visual than linguistic. Most users are pinning (or repinning) images or browsing, rather than writing comments. Popular categories of images on Pinterest include: arts and crafts, fashion, food, holidays and products.

Whilst viral images involve many of the principles of neuro design that we have looked at in previous chapters, they are a distinct class of images. They need to appeal to people in a different way than images that are meant to be merely attractive, desirable or aesthetically pleasing. They are also an image type where it is very clear to measure success: shares can be measured directly online. Whereas with images that are designed to create appeal, desire or be merely aesthetically pleasing, it is usually only possible to measure their success via setting up a dedicated research test. As computers get better at analysing images – both their content and style – there should be an increasingly sophisticated understanding of what makes an image viral.

Summary

- Visual memes are drawings or photos that get shared online thousands or millions of times because they tap into key things that our brains like.
- Designing an image to become viral is a different challenge from designing it to be beautiful or aesthetically pleasing. Virality and beauty are different, and beautiful images can sometimes be non-viral.
- Images that are synthetically generated or altered, are cartoonish/funny or contain animals are more likely to go viral.

- Emotional content is more likely to go viral than non-emotional content.
- High-intensity emotions are more viral than low-intensity ones.

Notes

1. Barrow, JD (2014) *100 Essential Things You Didn't Know You Didn't Know about Math and the Arts*, Norton, London.
2. Dawkins, R (1976) *The Selfish Gene*, Oxford University Press, Oxford.
3. Toffler, A (1981) *The Third Wave*, Pan, London.
4. Kelly, K (2016) *The Inevitable*, Viking, New York.
5. http://www.worldwideinterweb.com/8355-most-viral-photos-of-all-time-according-to-google/ (last accessed 25 August 2016).
6. Khosla, A, Das Sarma, A and Hamid, R (2014) What makes an image popular?, in *Proceedings of the 23rd International Conference on World Wide Web* (April), pp 867–76), ACM.
7. http://blogs.scientificamerican.com/scicurious-brain/you-want-that-well-i-want-it-too-the-neuroscience-of-mimetic-desire/ (last accessed 25 August 2016).
8. http://www.huffingtonpost.com/noah-kagan/why-content-goes-viral-wh_b_5492767.html (last accessed 25 August 2016).
9. Berger, J and Milkman, KL (2013) Emotion and virality: what makes online content go viral?, *GfK Marketing Intelligence Review*, 5 (1), pp.18–23.
10. Jung, CG (1991) *The Archetypes and the Collective Unconscious (Collected Works of C.G. Jung*, Routledge, London.
11. Deza, A and Parikh, D (2015) Understanding image virality, in *Proceedings of the IEEE Conference on Computer Vision and Pattern Recognition*, pp 1818–26.
12. Bakhshi, S, Shamma, DA and Gilbert, E (2014) Faces engage us: photos with faces attract more likes and comments on Instagram, in *Proceedings of the SIGCHI Conference on Human Factors in Computing Systems* (April), pp 965–74), ACM.
13. https://www.shopify.com/blog/9511075-the-science-of-retweets-10-steps-to-going-viral-on-twitter (last accessed 25 August 2016).
14. Gilbert, E, Bakhshi, S, Chang, S and Terveen, L (2013) I need to try this?: a statistical overview of Pinterest, in *Proceedings of the SIGCHI Conference on Human Factors in Computing Systems* (April), pp 2427–36, ACM.

Designing presentation slides 10

Figure 10.1 Bob began to realize that he would have won the $1 million investment if only he hadn't used Comic Sans...

In the mid-1980s a young Canadian film-maker walked into a Hollywood studio pitch-meeting and convinced the assembled executives to finance his movie idea just by drawing two lines on a blackboard. The film director was James Cameron and he'd written a sequel to the 1979 science-fiction horror movie *Alien*. The problem was that whilst the original was highly critically acclaimed it wasn't hugely profitable and the movie studio were not keen on bankrolling a sequel. The studio executives expected Cameron to arrive with a professional-looking slide show prepared, in which he would outline his plans for a sequel in detail. Instead he turned up without any slides, without even a piece of paper. He simply walked up to a chalkboard and wrote the word 'Aliens', then added two lines over the 'S', to turn it into 'Alien$', signalling that his plan was for a sequel that would be profitable. The meeting resulted in his movie being commissioned.[1]

Cameron's simple presentation contrasts with most presentations that use complex slides with lots of different elements. Since presentation software first became available to the mass market at the end of the 1980s, we have all had access to tools to create rich multimedia presentations. It is estimated that there are 20–30 million PowerPoint presentations given every day.[2] Words, photos, videos, clip-art, diagrams, shapes and an array of colours are all easily available. Yet the result is often presentation slides that are over-cluttered and suboptimal. We overuse the options available. The problem is often made worse with business presentations that are intended to double as reports. This results in information-dense slides that are designed to be read like a document, rather than being optimal for a talked-through presentation. If possible, it is better to create two versions: a minimalist one for presenting, and a more detailed one to share with people afterwards.

PowerPoint – the most popular slide presentation software – introduces invisible constraints that shape how presentations are structured. It obliges the user to break their presentation into linear, bite-sized (or slide-sized) sections, and encourages the use of hierarchical bullet-pointed lists. It can hide the complexity and rich interrelationships between information whilst giving the impression that each point has been covered. It forces audiences to focus only on the narrow range of information currently on display on one slide, constraining their ability to compare and contrast information across slides.

This format of design is so familiar to us that we don't question it. It doesn't leave much freedom for the structure of the message itself to dictate the structure of the design. In regular written text there is far less hierarchy, with information usually only broken down into headings and paragraphs. This particular problem of organizational structure is not present in all presentations. For example, the software Prezi, rather than using a linear sequence of slides, instead uses a large 'master' image on which all the information is organized, and then the presenter merely zooms in on each section as they present it.

On 16 January 2003 as the NASA space shuttle *Columbia* was launched from the Kennedy Space Center in Florida a suitcase-sized piece of foam insulation came loose and hit one of the wings. The launch was otherwise successful and the shuttle began its mission in orbit, but the ground-crew experts soon discovered the accident from video footage of the launch. The question was: how much risk was posed by the damage? Expert engineers rapidly compiled 28 PowerPoint slides summarizing their assessment of the risks. The slides used hierarchical bullet points to summarize their thoughts. However, their uncertainties and doubts were more confided to the lower-level bullets, whilst their higher-level bullets and executive summaries appeared to show a more optimistic assessment. The executive-level officials at NASA felt reassured and concluded that the shuttle was safe – and did not investigate it further. On 1 February, on its descent back to earth, the damage caused the shuttle's wing to gradually break apart, destroying the shuttle and killing all seven astronauts on board.

As part of the accident investigation, information design expert Edward Tufte was asked to analyse the PowerPoint slides, and pointed out how they could have been misleading. The official investigation board in their report wrote: 'The board views the endemic use of PowerPoint briefing slides instead of technical papers as an illustration of the problematic methods of technical communication at NASA.'[3]

Of course, most PowerPoint presentations don't involve such life-and-death decisions, yet with 20–30 million presentations being made per day globally there are probably a lot of important decisions being made and resources allocated via this mode of communication. PowerPoint bullet-pointed hierarchies may obscure people's messages by overemphasizing some points and hiding others. Tufte believes that PowerPoint is ideally used for presenting (relatively low-resolution) images and videos, but cautions against using built-in templates that pre-structure your information for you, and hierarchical bullet points.[4]

Allowing the audience to follow your message

Presentations usually need to do several things: engage the audience, persuade them, and give them information in a way that helps them remember it. Yet presentations need to do this in a way that accurately communicates the information and does not confuse or mislead. Good presentation slides that are designed in a way that is optimal for the way our brains work can help with these goals.

Presenting information from slides benefits from thoughtful preparation. Slide presentations, if not done carefully, can be an inefficient mode of communication. For example, when reading a book, people can proceed at their own pace, and if they don't understand something or have forgotten something they can pause or look back. This is not always possible when people are viewing a presentation. Equally, it is not always easy for presentation viewers to stop the presenter and ask them to explain a point that isn't clear. Viewers are forced to take in information at the presenter's pace. Therefore it is important to design slides in a way that will help people to follow your message.

In order for the audience to follow your message, they need to do several things: pay attention to the most important messages, understand how information on each slide connects together, and link together information across slides. For each of these there are some neuro design principles that can help, as set out below.

1. Pay attention to the most important elements on the slide

Slides can contain a lot of information, so in order for viewers to make sense of it, it is important to indicate which bits are most important. This is where visual salience can be useful. Decide which element on the slide is most important for people to see and think about first – and make sure it has the highest visual salience. This can be done by:

- making it the largest element;
- isolating it with plenty of space around it;
- putting a high-contrast box around it;
- using a colour that contrasts strongly with the background and the rest of the slide.

If you use a background colour or pattern, ensure it is not too salient – it might overwhelm the other visual elements on the slide.

The ideal way to convey importance is with both salience and a large, prominent title. We use size as a way to decide the hierarchy of importance, with larger things taking prominence over smaller things. So larger elements are both more likely to be seen, and considered important.

Isolating an element visually also helps draw attention to it. The von Restorff effect is the tendency for the most distinctive element within a design or a list of items to be the one that is best remembered. Isolating an element marks it out as being different and therefore something that people should pay attention to.

Finally, introduce the most important idea first. If you have a key idea that you want people to remember after they have left your presentation, make it the subject of your first slide. This will help make it more memorable.

2. Understand how the different elements on the slide connect together

Viewers will more easily connect items together if they are positioned closely together than if they are merely connected in some other way (such as sharing the same colour or shape). Consider whether you can position words around the slide to show their relationship (ie rather than just have a list of words in a vertical column, could they be arranged or grouped?).

Good graph design

Charts can be a good way to portray information visually. We have a good intuitive grasp of the way that size and importance relate to visual features such as size and height. Taller bars mean more; bigger segments on a pie graph mean more; a line moving upwards means an increase. Charts are a metaphoric way to convey information but they are close enough to how we understand size, quantity, increases and decreases in the real world that they make sense.

Many slide presentations include charts, and these are easy to produce in software such as Microsoft Office. However, the ease with which they can be created sometimes undermines the ability of the chart to properly convey information. Make sure you match the right chart design to the information you are trying to depict.

It is important to consider if you are dealing with continuous or discrete data. Continuous data is a measurement of something that is changing over space or time, such as temperature movements over the course of a year. Discrete data is individual measurements of different things, such as sales of different products:

▶

- **Line charts**: use these when you need to show a trend along a continuum, such as over time. Crossing lines on a line chart are the ideal way to depict interactions of measurements.
- **Bar or column charts**: use these for specific comparisons between discrete scores. Here the heights of the bars effectively convey the sizes of the different quantities. If there is a 'goal' figure that each quantity is trying to reach (such as a maximum score or sales target) consider sideways bars, moving towards the right, as this gives the feeling of a race, with the higher-value bars out in front, closer to the finish line.
- **Pie charts**: these are best used when you need to show how the parts make up the whole of something, such as a series of percentages that add up to total 100 per cent.

Another good chart design practice can be to consider first ordering or ranking your data before you chart it: for example, ordering scores from highest to lowest in a bar chart. This can help make the data more intuitively clear.

Lastly, the evidence for the benefits of 3D graphs over 2D is mixed to negative. It is probably usually best to stick with 2D.[5]

Increase the processing fluency of slides. By keeping slides as simple as possible, you will be eating up the minimal amount of viewers' mental effort, making it easier for them to understand and remember your presentation. In particular, try to keep text to a minimum. The more text there is, the more viewers will feel obliged to read it all and miss what you are saying. Even worse is the possibility that they will feel the need to write it all down, in which case they may completely miss what you're saying.

Try to make each element on the slide support the other elements by giving them a cohesive feel: each element should feel like it belongs together and is supporting the same message. For example, use fonts that are appropriate to the subject at hand. If it is a fun or light-hearted presentation feel free to use a more playful font, but if it is a presentation intended to impress with a more weighty message be careful of your font choice. When the scientists at CERN made the historic scientific announcement of the discovery of the Higgs boson particle, they were criticized for using Comic Sans on their slides.[6] This font looks childish, undermining the import of their message.

In one study that examined 10 popular fonts, the researchers recommend using Gill Sans for slide presentations.[7] They asked participants to rate each font for how comfortable it was to read, and how professional, interesting and attractive it looked. Overall there was no significant difference between Serif and Sans Serif fonts on ease of reading, interest and attractiveness. However, the Sans Serif fonts were significantly more likely to be rated as professional looking (although this was driven by low rating for two of the five Serif fonts: Garamond and Lubalin Graph Bk), although the font with the highest rating on looking professional was a Serif font: Times New Roman (with the Sans Serif font Tahoma receiving the second-highest rating). Overall, across all four measures, Gill Sans was the strongest performer, hence their recommendation.

Keep text easier to see for anyone in the audience who may be colour blind by avoiding red on green or blue on yellow. The larger the audience, the more likely that some viewers will be colour blind.

3. Remember information from previous slides in order to link ideas together

Whilst a presentation tells a story, or makes a particular argument, it is split up across multiple slides that people see one after the other. As soon as each new slide comes up, the previous one disappears. This means that viewers have to hold in mind information from previous slides in order to connect it to the information in the current, visible slide. As the presenter you implicitly understand the connection between the points in each slide – you have seen the overview of the whole presentation. However, this is not always clear to viewers, so it can be important to help them follow these connections.

As we have already seen, our short-term memory is limited. We can only hold a short list of ideas in mind at once. If too many new ideas are added to that list, the older ones drop off. There are essentially two ways to help viewers with this problem. First, you can give them a visual reminder of information from previous slides so that it is still easy for them to keep it in mind. One way of doing this might be to include a thumbnail image of the previous slide in the bottom corner of each slide. This only really works, however, if the gist of that slide is still clearly viewable at a smaller size. Text might not work that well, but images or information that is organized spatially might. This strategy could either be used on every slide, or just selectively when you think there may be a difficult concept for viewers to understand that would benefit from a reminder of the previous slide. The other way to help people's limited short-term memory is by chunking information together for them. Whilst people can only hold in mind a handful of items at once, those items can each be 'chunks' of closely connected items. By grouping concepts together you can therefore maximize people's ability to hold on to information, and hence understand your presentation. For this reason, don't include lists (eg bullet-point lists) of more than about four items, and don't include more than two lines per item.

Also, don't use arbitrary changes in style. People expect a change in style – such as changes in font colours, background colours or type of font – to connote changes in meaning. Changing them randomly can therefore confuse viewers.

The doorway effect may apply to transitions from one slide to another. In Chapter 8 we looked at the doorway effect on memory: the tendency for scene changes (such as when you walk through a door into another room) to make it harder to remember information from the earlier scene. A slide transition might make it harder for people to retain in mind the information from the last, or earlier, slides that are now no longer visible.

The corollary to this is that if you are changing the subject within a presentation, you can use the doorway effect to give closure to the previous topic before introducing the new one. Add in a pause, both in your verbal delivery but also visually: for example, by including a slide that summarizes the previous topic, or that is mainly blank but just includes the title of the next topic.

In a survey of people who regularly view slide presentations the most frequently cited annoying feature of presentations was too much information being included on each slide.[8] In a follow-up study they showed participants pairs of slides in which one slide was better designed (according to neuro design principles) than the other and tested whether they could spot the better slide, and provide an explanation for why it was better. People were most likely to spot when slides had too much information on them. They were least likely to identify mismatches between the design choices and the information being conveyed. For example, when the wrong style of graph was chosen, or when the style of graphics used was not appropriate for the messages being conveyed. Participants in the survey were equally unlikely to spot the mistake of slides with bad visual salience. This is interesting because whilst we know that visual salience has a measurable effect, this survey suggests that viewers are not always aware of the impact of bad salience on design. The authors of the study concluded that: 'even if viewers do not notice a violation (of a neuro design principle) they could still be affected by it'.

Here are some extra tips for minimizing the mental effort required for viewers to decode slides, so it is easier for them to follow your presentation:

- **Image on the left, text on the right**
  As we saw in Chapter 3, there is evidence that people prefer designs with the text positioned to the right of any images. This is because we process images and text better when they are positioned this way. It gives a design, like a presentation slide, better processing fluency. In other words, it is slightly easier for viewers to take in the information.

  Another related effect that we saw in Chapter 3 is pseudoneglect: that we naturally overestimate visuals presented in the left of our gaze over those in the right. The left side of your slides will receive greater visual attention. It's the natural place to put images. The text on the right does not need so much visual attention as it is usually just repeating some of the words you will already be saying out loud.
- **Audiences cannot read words and listen at the same time**
  If your slides are full of words people will either read them and ignore what you are saying, or listen to what you are saying but not read the words. We cannot do both. We are able to take in one 'stream' of words at a time, not two. For example, it is easier to listen to instrumental music and read at the same time, than listen to a verbal conversation and read at the same time. If we have to listen to and read words simultaneously, what we are actually doing is rapidly switching our attention back and forth between the two. Psychologists call this 'divided attention', and it is effectively weaker than if you concentrate your attention on one verbal 'stream' of information.

  However, people can look at an image and process it whilst listening to what you are saying – adding another reason why it is best to use the parallel channels of both images and words.
- **The 10-minute rule**
  Evidence suggests that people can watch a presentation for around 7–10 minutes before their attention starts to wander. Presentations can be relatively 'low-stimulation' compared to things like TV shows and movies, so the challenge is to keep people engaged for extended periods of time.

  If your presentation is longer than 10 minutes it is therefore a good idea to inject some variety 10 minutes into it, and every 10 minutes after, in order to keep people's attention from flagging. This can be done in a number of ways:

  - show a video;

- get the audience engaged in something interactive;
- show a surprising image.

Anything that can help change the pace, or inject some variety at around the 10-minute point can be useful.

- **White might not be the best background colour**
  Many, perhaps most, presentations use slides with a white background. If your projector or screen is powerful, and if the room is dark, white might not be the best background colour. This is because large areas of bright white can fatigue people's eyes.

Visual learning

Our understanding of visuals predates our use of language. Dreams, produced by our non-conscious minds, are more visual than linguistic. Research shows that people learn better from images than words. Images are processed more directly by our brains, whereas words need first to be processed visually, with our visual brain recognizing each letter, before then interpreting those letter-images into words, then trying to convert the flow of those words into meanings. 'Your soul,' said Aristotle, 'never thinks without a picture.'

We are all familiar with the ways that our eyes provide a parallel channel of information to that from our ears. Even when listening to someone talk, a lot of the information that we glean from them is from their posture, their facial expressions, eye contact and so on. The very meaning of what they are saying can be modified or changed depending on these things. If we can see visuals that support and add to the verbal stream of information we get a richer, deeper experience. Similarly, in graphic novels we may be reading the text, but the accompanying imagery adds a distinct layer of meanings that can make the story clearer. In a well-designed graphic novel the visuals will be adding to the text, not merely saying the same thing. In a lot of presentations the slides can be redundant, as the presenter just reads the text from the slides. Text-only slides are an inefficient approach if the presenter is simply reading from the slides anyway.

For example, one study flashed up more than 2,000 images to volunteers, with just 10 seconds per image. Several days later they could correctly recognize which images they had seen with 90 per cent accuracy. Even a year later they could recognize them with an accuracy of around 60 per cent.[9]

The trouble with treating presentations as a linguistic medium is that slides cannot contain as much text as pages in a book or report, so text gets summarized – it is a compromise. Visuals allow us to compress more complex ideas into a format that can work on slides. As we saw in Chapter 3, images can convey many different ideas, even if they are graphically very simple (ie images with high 'propositional density'). This makes them easy to look at, but rich in meaning. Research on memory also shows that the more deeply we process information, the more likely we are to subsequently remember it. By using both channels – visual and linguistic – people are more likely to process the information more deeply.

Other research shows that we process the flow of images within the panels of a comic in a similar way to how we process sentences.[10] Graphic novels are increasingly popular worldwide, perhaps an indication that people prefer taking in information visually, or at least visually as well as linguistically.

Adding visuals along with the text, especially if the visuals add a new layer of extra information or insight, will help make your message more memorable. Ever since ancient Roman times people have used imagery to make information more memorable. Competitive memory champions use techniques involving turning dry information into images. For example, memorizing the long sequences of numbers in pi, memory champions will turn those numbers into visual images and link them together in a story. Our brain has evolved to handle sequences of images well.

There is evidence to support the idea that adding graphics makes information more memorable. In one study, participants were split into two groups. One read a section from a textbook about business management. The other group read a graphic novel that covered the same material but with graphics as well as text. Those who had read the graphic novel were later better able to recognize quotes from the material than those who had read the text-only version.[11]

Steve Jobs, in his famous Apple presentations where new products were announced and launched, used some of these neuro design principles. He would keep the number of words on a slide to a minimum, helping to keep it easy for viewers to follow what he was saying. Further, the phrases used on slides would be honed into snappy, clear and memorable lines that were headline and Twitter-ready. In other words, journalists and viewers could easily directly take these sentences and use them as their own headlines or tweets to summarize the key message of the presentation. Again, this style minimized the effort viewers needed to exert in order to follow-through on the information he was presenting: in this case remember it, and write about it.

The way that presentation slides are designed can make the difference between a good and bad presentation. It is also a form of courtesy to your audience to put in the preparation effort to make your slides as easy as possible for them to understand and absorb. It can also ensure that your message is communicated as intended, and not misunderstood.

The power of visual stories in presentations

Christopher John Payne is a marketing consultant who helps a broad range of clients, from accountants to dating coaches, to improve how they communicate what they know. He says:

> *Webinars – online presentations – have become the key way to sell, and what really sets the successful presentations apart from those that are merely 'okay' is having lots of visual stories. I have become an evangelist for trying to log a personal story every day, and taking photos to illustrate them. Evolutionary psychologists would back up the power that stories have on our brains. Our ancestors were telling each other stories every night around the campfire for millennia. They help to make your points more interesting, memorable and convincing. We all have more stories than we realize, so the habit of logging them, along with photos to illustrate them, is invaluable when it comes to putting together presentation slides. Pop a photo into the slide to accompany your story and you will massively increase audience engagement. In fact, I'm so dedicated to capturing story photos that when I was rushed into hospital in an ambulance recently, despite being in tremendous pain I gave one of the paramedics travelling with me my phone to take a photo of me on the stretcher as we hurtled along the road. I knew I'd one day find a way to capitalize on the story in a future presentation!*

Summary

- Think very carefully about using nested hierarchies of bullet points in presentations, especially ones that are helping people to make important decisions. This format can be misleading.
- Neuro design ideas can help design slides that draw attention to the most important elements, helping to ensure that you get your message across.

- The main problem with most presentations is slides containing too much information. People are more likely to pay attention, follow your points and remember them if you design your slides with the minimal amount of text.
- Combining images with text on every slide can help convey more information easily for people viewing the presentation.
- By using neuro design principles you can minimize the amount of effort needed by viewers to follow your message, making it easier for them to connect points across slides.

Notes

1 This story was related by Gordon Carroll, the executive producer of *Aliens*, in: Obst, L (1997) *Hello, He Lied: And other truths from the Hollywood trenches*, Broadway Books, New York.

2 Alley, M and Neeley, KA (2005) Rethinking the design of presentation slides: a case for sentence headlines and visual evidence, *Technical Communication*, 52 (4), pp 417–26.

3 Columbia Accident Investigation Board, Report, vol. 1 (August 2003), p 191.

4 The cognitive style of PowerPoint: http://users.ha.uth.gr/tgd/pt0501/09/Tufte.pdf (last accessed 25 August 2016).

5 Tversky, B, Morrison, JB and Betrancourt, M (2002) Animation: can it facilitate?, *International Journal of Human–Computer Studies*, 57 (4), pp 247–62.

6 http://www.theverge.com/2012/7/4/3136652/cern-scientists-comic-sans-higgs-boson_(last accessed 25 August 2016).

7 Mackiewicz, J (2007) Audience perceptions of fonts in projected PowerPoint text slides, *Technical Communication*, **54** (3), pp 295–307.

8 Kosslyn, SM, Kievit, RA, Russell, AG and Shephard, JM (2012) PowerPoint® presentation flaws and failures: a psychological analysis, *Frontiers in Psychology*, **3**, p 230.

9 Medina, J (2014) *Brain Rules*, Pear Press, Seattle.

10 http://discovermagazine.com/2012/dec/29-the-charlie-brown-effect (last accessed 25 August 2016).

11 Short, JC, Randolph-Seng, B and McKenny, AF (2013) Graphic presentation: an empirical examination of the graphic novel approach to communicate business concepts, *Business Communication Quarterly*, **76** (3), pp 273–303.

Conducting neuro design research

11

Figure 11.1 Some of the key neuro design research tools (from top, left to right: EEG, A/B testing, eye tracking, implicit response, facial action coding and fMRI)

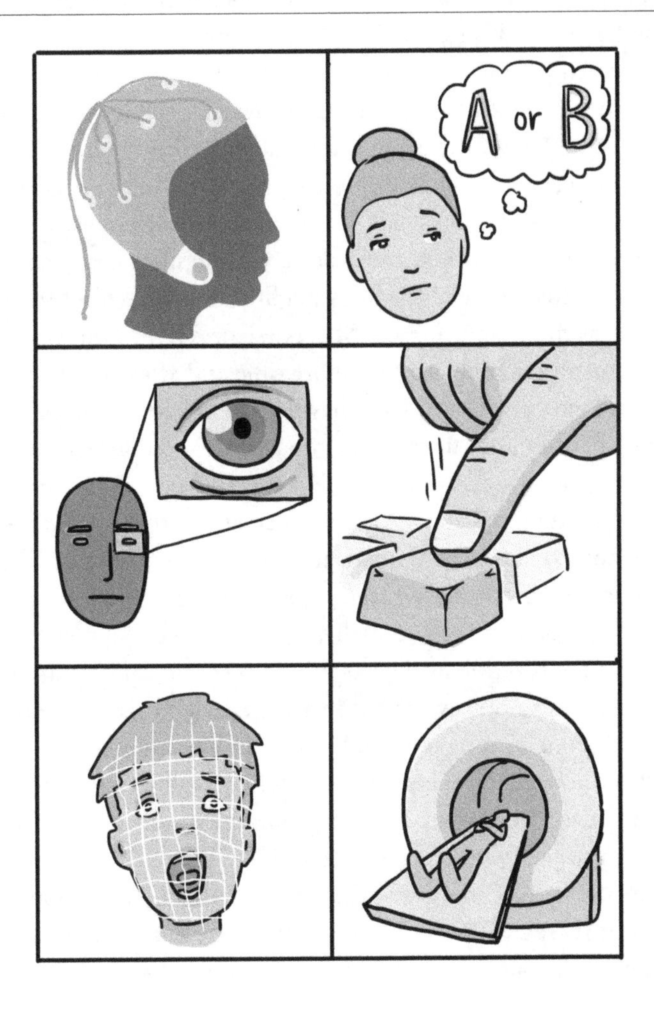

Dilbert is one of the most popular comic strips in the world, published in 2,000 newspapers across 65 countries. Over 20 million Dilbert books and calendars have been sold. The cartoon features the eponymous Silicon Valley office worker and offers satirical observations on some of the more absurd and irrational aspects of the contemporary workplace. It originally appeared in 1989 in a handful of newspapers but did not begin to really take off until 1993 when Scott Adams, its creator, added one small feature at the bottom of each strip: his e-mail address. Previously he had only received direct feedback on Dilbert from friends and colleagues, but by adding his e-mail address and letting readers contact him directly (something unusual at the time) he suddenly got loads of feedback from his actual end customers. Whilst his friends and colleagues kept their feedback positive, via e-mail his readers were not scared to say what they didn't like. The biggest insight was that most people preferred seeing Dilbert in the office more than at home. When Adams implemented this feedback, Dilbert's popularity began to soar.[1]

Of course, simply inviting users to comment on a website or product is not always enough. With cartoon strips the design itself is the product. They are (if well done) highly engaging and people like talking about them. Therefore adding in the e-mail address of the artist may be more likely to yield responses on a cartoon strip than on most other designs. Nevertheless, the idea of testing and getting feedback on your designs is a good one.

As we have already seen, designers might not always react to designs in the same way as the average person. Many designers have spent years studying art and design, and they spend a lot more time and effort in their daily lives looking at and judging designs. They may have even been attracted to working as a designer because they have a very good 'eye' for design. This 'eyeballing' of their own design work as they develop it is tremendously useful, and the designer's own intuition is still the engine of creativity that drives good design. However, what may be most aesthetically pleasing to the designer might not be the optimal design for the average population, who might be less visually and artistically sophisticated. Equally, it is not always known what will trigger the desired reaction in viewers. Without testing and research it can be hard to know, for example, what it is in a pack design that will be the biggest trigger for someone picking it up and buying it. We can predict features that will help it get noticed (visual saliency), or will make it attractive (processing fluency). However, consumers could be prioritizing any number of different features of a product when they decide to buy. With a food product it may be the taste, the texture, the perceived quality and so on. Without knowing the priority of these features it is possible to create a pack design that might be beautiful and attractive, but does not work as hard as it might in creating consumer desire. Even if you do have a reasonable idea of what the most important triggers might be, choosing the best content or subject matter to evoke the desired response can be a challenge that research can help with.

This is not to underestimate the importance of the neuro design principles we've covered so far in this book, but to say that the ultimate way to optimize a design is to both use these principles and then to test the design. As mentioned in Chapter 1, neuro design, including neuro design testing, works best to augment designers' intuition, not replace it. So the ultimate mix is:

A designer's own creative ideas + neuro design principles + testing

Of course, for much of design work there is not always enough time or money to conduct tests. Nevertheless, there are some reasonably cheap and quick ways to test. Testing does not even have to wait until the finished design is finalized, it can be used throughout the design process.

The benefits of actually testing designs are similar to those for testing new drug medications. A drug may work in theory – the biochemistry behind its design may make sense at a theoretical level – and it may appear to work when tested on a small number of people. Yet it is only when a wide-scale controlled clinical trial is conducted that we know for sure. It might be that the drug – which appeared to work well when tested on a few people – has unexpected side effects when tested widely.

The first principle of neuro testing is that people can only react to what is in front of them. They cannot do the creative work for you, and tell you what their ideal design should look like. The neuro design principles that govern much of our reactions to design are largely non-conscious, therefore we cannot sense them occurring in our brains and report them back to researchers. Equally, consumers are not creatives, it is not their job to come up with the creative concepts and compositions of designs. Therefore it is generally best to have more than one design option to put in front of people for testing. This allows you to compare reactions to actual designs, rather than speculate about which design idea might fare best.

Perhaps the best way to test is a real world A/B test, whereby at least two different design options are placed out into the real world (eg on a website), with a random selection of people seeing one or the other. The main advantage to this method is that the results are actual real-world performance. The main weakness is that even if you find a clear winner amongst your designs, you will not necessarily know exactly why it won. You can make guesses, but you are not necessarily generating new insights into what works with your customers. Another weakness of A/B testing is that it is not always possible. Whilst it can be cheap to do online, in the real world with physically printed designs the extra production costs and time schedules can make it impractical.

Average versus polarizing results

When looking at the results from testing it is often a good idea to look at the spread – or distribution – of results as well as the average. An average response that is slightly positive might be composed of most people having a mildly positive response to your design, or by some people loving it and some not. Not every design needs to appeal to the mass market, to everyone. It might be better for your designs to gain an intensely positive response amongst a select minority of people than for it to be merely 'okay' for most.

Controlling for other factors

In any scientific research it is important to try to ensure that you can be sure of what created your results. This means that if you are testing the different reactions that might be elicited by different design options for a webpage, package or product design you need to make sure that it is only the design that changed between tests, otherwise you will not know if it was some other factor that caused the difference. Here are some of the main factors you should control for:

Who you are testing it on

The key rule is to make sure that you test all the images (or videos) that you are comparing on the same group of people. If you cannot show all the images to exactly the same people, in other words if you need to recruit more than one group, then make every effort to ensure those groups don't differ in any way that could create different responses. For example, typically in market research tests, the researchers use a screening questionnaire to qualify people for a test. These questions might cover things like the demographic details of potential participants, but also things like whether they currently buy your brand/product or use your website. These are the key things that could affect how they react to your designs. Make sure you use the same qualifying questions on any groups you recruit.

The content of the designs

If you are testing the designs of two different webpages, but there is different content on those pages, you ultimately will not know for sure if it was the content or the design differences that drove people's reactions. Similarly, if you were testing package designs for two food or drink products, but they were also two different flavour variants, you will not know if it was the flavour or the design that drove the difference. Try to keep the content the same between different designs.

The way you run the test

If you are trying to compare results across different tests, but the tests were conducted differently, again, you cannot be sure if it was the designs or the test method that drove responses. Keep the exact same test procedure across all tests that you are hoping to compare results from.

The importance of context

In the real world we react to designs within a particular context, whether it is shopping online or in a supermarket, browsing a magazine or seeing a billboard on the streets. The same design may perform differently depending on the context. For example, a package design on a supermarket has a lot of competing packs that it needs to perform well against.

There are three main ways that you should consider the context of the material you are testing:

- what you tell participants at the beginning of the test;
- what mix of images each participant sees;
- how they see the images.

What you tell participants at the beginning of the test

What you tell participants before you show them your designs can influence how they react to them. Usually it is best to say nothing, other than ask them to look at the designs. However, if the designs are ambiguous in some way, you might want to explain what they are. For example, if the designs are for a new innovative product or website, you might want to explain a bit about it first.

What mix of images each participant sees

If you are showing study participants a range of images or videos, they will inevitably compare them to each other. First, this means that it is important that you randomize the order in which people see them, so that the particular order or sequence of images seen is not itself driving the overall results. This also means that you need to consider how the mix of images seen might affect results. For example, if you have a batch of images that you need to test amongst two or more groups (eg because there are too many images to show to one group), then how you organize those batches is worth some thought. If you are testing package designs for two different flavours of food, try to batch all the packs of one flavour with one group, and all the other flavour with the other group. Similarly, try to avoid an 'odd one out' effect with images, ie don't test a group of images that are very similar mixed in with one that is very different. The different one may receive an unusual response just because it stands out against the rest.

How they see the images

You can display designs in a test either in isolation or in context. Isolation would be the design simply on its own, or against a plain background. In context might mean displaying a pack design embedded in a shelf display (perhaps against its competitors), or a billboard poster design embedded on a photo of a billboard in a street photo. It might seem like testing designs in context is the superior technique. After all, that is how they will be seen in the real world. However, the challenge is that any particular photo or re-creation of a context is showing a very specific example of the real world. For instance, not all shelf displays look the same in all supermarkets. Your results might be driven by placing two particular packs next to one another, but in many supermarkets they might not be seen together. Or if you embed your poster design into a photo of a street billboard, there might be other things going on in that street photo that are affecting people's responses.

One solution to this is to provide a visual 'hint' of the context, by either cropping the image quite tightly around your design, with the background context just shown around the edges of the image. Alternatively you can make the background details more indistinct by applying some form of filter (such as a greyscale or blurring filter) so that people get the general impression of the background, but cannot see the details as clearly as they can see your design.

The new research tools

Matching the right tools to the right research question is important. There is now a range of neuro research services that you can commission to test your images on. Many of these work online, and hence are quicker and cheaper than physically bringing test respondents into a special location for testing. Some even have 'self-service' options, where you subscribe to an online portal and set up tests yourself (usually after undergoing some training from the provider).

Given below are some of the main methods available.

Online eye tracking

Eye-tracking technology uses cameras pointed at a person's eyes whilst they are looking at a screen or display, in order to monitor their eye movements and hence what they are looking at from second to second. The technology has been around for decades, but usually required a dedicated lab set-up, or at least bringing volunteers into a facility where the cameras are set up. In recent years it has become possible to do eye tracking online using people's own home-computer webcams. Conducting eye tracking online has the benefits of being cheaper and faster. It is more costly to pay people to physically travel to a location to be measured via a dedicated eye-tracking camera than to get them to sit at home and do a quick online test.

The procedure for online eye tracking is very simple. People are recruited online to take part in the test and they are sent to a link. It explains that they will need to have a working webcam and asks for their permission to access it. They then have to make sure the light is on in their room, and they are asked to keep their head still whilst they look at the screen, and try to move only their eyes. They then typically perform a calibration test, to help the system track where they are looking on their screen. For example, there might be a moving dot that systematically moves to different positions all around the screen as the person is asked to follow it with their eyes. The system is then able to correlate different movements of the person's eyes, as captured via the webcam video, with where it knows the dot is at that moment in time on the screen. Finally, the images that the researcher wants to test are displayed on the screen in random order (to ensure that the reactions of the group, overall, are not due to the order that the images were shown), typically just for around 5–10 seconds per image.

For an eye-tracking study you typically need around 20–30 good recordings. A good recording means the person has kept their head relatively still, the lighting conditions are good, and in particular the person's face is at least reasonably well lit. In a lab or controlled testing environment it is easier to control these things than when a person is doing the test themselves at home on their own computer. This means that with online eye tracking you often have to run up to four or five people through the test for every one good recording you capture. It's still cheaper though, as it is much cheaper to recruit test participants online than for a physical location study.

The benefits of eye tracking as a methodology

It is non-invasive: the person does not have to be in an unusual or artificial environment, they are just sitting in front of a screen. They don't have any sensors or unusual technical equipment around them or attached to them.

It is fairly intuitive to understand. The outputs show where people looked on the screen, and this can be visualized in two fairly intuitive ways. First, a gaze-plot. This shows a series of circles (or squares/rectangles) overlaid on the image, one circle over each image element that received a lot of attention. Typically the size of the circle represents the amount of attention that area received. Then the most frequent order in which those areas were looked at can be depicted by giving each circle a number and connecting them with lines or arrows. This gives some clue as to the priority of what people are drawn to look at.

Second, heat maps, or their reverse version, fog maps. This is a colour overlay over the original image in which zones that are receiving information have a cloud of warm colours. The way this is most often depicted is with areas that receive moderate attention being covered in yellows or oranges, and areas receiving a lot of attention in red. In other words: the warmer the colour, the more the attention. The weakness of this visualization is that if there are yellows, oranges or reds in the image itself it can be difficult to tell which colours are part of the image, and which are part of the heat map. This problem is overcome with the opposite to a heat map: the fog map. Fog maps obscure with a hazy fog the areas of the image that got the least attention, whilst areas that got a lot of attention are left clear. This shows you instantly and intuitively what parts of the image are 'burning through' to people's gaze.

Design questions that eye tracking is suited to answer

- Is a particular element within a design getting seen?
- What do people look at first?
- What do people look at most?
- Do people seem confused in their gaze pattern on the design?

Implicit response measures

Implicit response measures are a group of computer tests that measure the connection between an image and a concept, idea or feeling. For example, if you wanted to test several possible designs to see which is best at evoking a particular reaction, this would be a good technique.

It is based on the concept of priming. Whenever we see something, even for a split second, all the associations we have with it become more available to our brains. In other words we are primed to recognize those associated things more readily.

The tests can be structured differently but they all work by measuring the reaction speeds of people clicking keys on a keyboard or swiping on a touchscreen. Participants are asked to perform a simple categorization task, whereby words or images flash up on the screen one after another and they have to categorize them correctly into one of two 'buckets'. For example, there could be words that are emotionally positive or negative, and the task is to click one key if it's a positive word, another key if it's a negative word. It is like a very simple video game. However, before each word or image to categorize appears, it will be preceded by a brief glimpse of another word or image. This is called the prime, as it is displayed in order to trigger or prime an association. In the example of the task of categorizing positive and negative words, if a picture of an attractive beach was briefly shown before a positive word, the positive associations primed by the beach image would likely then speed up the person's ability to categorize a positive word that appears immediately after it, but it might slow down their categorization of a negative word as there would be a momentary clash of concepts in the person's mind. Conversely, if the positive word was preceded by an image with negative connotations, such as a graveyard for example, it would likely slow down people's categorization of a subsequent positive word, and speed them up on the categorization of a negative word. Analysis is then done on the comparative reaction speeds of each word/image pairing in order to see which are more closely associated in people's minds.

An important feature of implicit tests is that they are measuring reactions indirectly. They are not asking the person directly: how closely do you associate this image with positive feelings? And they are not asking the person to directly categorize a picture, as you might do in a traditional questionnaire survey. Instead, the categorization task has objectively right and wrong answers. If someone gives the wrong answer, the system can ask them to try again. This means that the person is forced to stay focused on the task, they cannot 'cheat' by just repeatedly pressing the same key or giving the same answer as they potentially could in a standard online survey.

As with eye tracking, implicit response measures can be done online. Indeed, it is possible to combine the two tests, one after the other, within the same online testing session.

One limitation of these tests is that you cannot get a moment-by-moment read on experiences like watching a video, or moving through a website. You get an overall read for the experience.

Design questions that implicit response measures are suited to answer

- Does this design evoke the emotions that it was intended to?
- Is the intuitive reaction to this design positive or negative?
- What are the different associations that people have with your images?

Designers using neuro design

London-based Saddington Baynes are a creative production company who have adopted neuro design research and thinking in their creation of images.

When designing things like car ads for magazines or posters, they have discovered that small details can have a big impact on what the image conveys. For example, changing the lighting, colours, camera lens or angles, background locations or the orientation of the car can all shift perceptions.

My colleague Thom Noble and I helped them to set up a regular neuro-testing capability that they have embedded into their design production process. Beginning with a research and development phase, we helped them to understand the impact of these design changes by systematically testing different sets of car images. In each set we systematically only changed one thing (such as the colour of the car, or the angle it was shot from), and measured the effect this had on both the emotional appeal of the image and the types of attributes that viewers associated with the image (such as its ability to convey style or excitement). This provided a series of hypotheses to guide them in how to compose future images, based on the effects they want to achieve on each image.

The system has now been refined to a practical test-and-learn tool for the designers. If they want to double-check that a new design is working well, or to test which of several design choices will work best for the goals of a campaign, they can run the images through their system (which tests on several hundred people online, using a neuro research method called implicit response testing).

Facial action coding

There are six facial expressions of emotion that seem to be innate to everyone, no matter where they were born: happiness, surprise, sadness, fear, disgust and anger. Research across different cultures – even those not exposed to media like TV (where people could have learned to associate different expressions with different emotions) shows that they are innate. We also know that at least some expressions of emotions on the face are innate rather than learned through observing others – as blind people also make them.[2]

Facial action coding – or FAC – can recognize the expression of any of these emotions, moment by moment, by taking camera footage of a face and figuring out muscle movements that control the display of each distinct emotion.

One weakness of FAC is that it is limited to the universal emotions. Many images and videos are trying to evoke different emotions. Nevertheless, if you do have an image or video that is designed to evoke one of these emotions, particularly if it is strong enough to expect to generate actual facial reactions, it can be a good measure. It is particularly useful with measuring reactions to videos.

As with online eye tracking, facial action coding can now be done over the web, via users' own computer webcams – again, bringing the benefit of reduced costs over having to bring people into a lab or research facility.

Design questions that FAC is suited to answer

- Are people smiling at a joke in your video ad?
- Does the surprising moment in your video genuinely elicit surprise from viewers?
- Does a scary horror movie trailer evoke fear in viewers?

EEG/fMRI

EEG and fMRI are more complex, specialist and expensive techniques that use technical equipment to measure brain activity directly.

fMRI – or functional magnetic resonance imaging – is a large chamber in which people lie whilst the machine scans blood movement within their brains as a measurement of which brain regions are 'active'. From this data the analysts can deduce the person's reactions to things like images and videos. Many of the major findings from academic neuroaesthetics come from fMRI research.

EEG – or electroencephalography – involves placing a series of sensors over a person's head (sometimes with a water-based gel on them), usually positioned in a cap, in order to measure patterns of electrical activity emanating from the top part of their brain, the cerebral cortex. EEG can be used to measure a number of types of responses, for example:

- Attention: EEG is particularly good at measuring the amount of attention that a person is paying to what they are looking at. Their eyes may be looking at an image or video, but are they really engaged with it? EEG can tell.
- Cognitive load: similar to attention, cognitive load is a measure of how hard the person's brain is working to decode what they are seeing.
- Emotional motivation: does the person feel emotionally drawn into or repelled from what they are looking at? EEG can measure this by monitoring different patterns of activity in the left and right hemispheres of the frontal cortex.

An alternative version of EEG is SST – or steady-state topography. This is an enhancement of standard EEG recording. The person wears a cap with EEG sensors, but the cap also introduces a dim flickering light. This light sets up particular, predictable frequencies in the person's brain, as their brain reacts to it – a little like a person humming along to a tune. Then, when different areas of their brain become active, or more busy, in attending to what they are seeing, this copying frequency starts to deviate (like the person's humming slowing down or getting more quiet as they concentrate on something).

fMRI and EEG are expensive techniques, and therefore most appropriate for projects with large research budgets. Research techniques can also be combined. For example, eye tracking can be done at the same time you are measuring a person's facial responses with FAC, or their brain responses with EEG.

Research questions that EEG, SST and fMRI are suited to answer

- Which parts of your video do viewers pay most attention to?
- How can a video be re-edited to improve it?
- How emotionally engaging is your video?

Keeping up with new research findings

Academic psychology and neuroscience papers can seem impenetrable. They are often written in what seems like insider 'jargon' and filled with complex statistics. However, if you are motivated and read them carefully they can be understood, even if you don't have an educational background of reading them. Here are some tips.

First, it is useful to understand how neuroscience papers are structured. Most papers follow a similar structure; if you know the main sections in this structure it makes understanding papers easier:

- **The abstract**
  Most papers begin with a section called an abstract. This is a summary (usually just a paragraph or two) at the beginning of the paper that covers the main research question, what was done to research it, what they found and what they conclude. If you read this section you can usually judge whether the paper is of relevance to you.
- **Keywords**
  Immediately under the abstract are usually keywords that the paper covers. These are like descriptive tags that define the topic. They can be useful as they often show the scientific terms that researchers of that topic are using. If you want to research a particular area of design you need to know the inside jargon. You might be using terms that are different from those of the research community. For example, you might want to find research on visual stand-out. If you didn't know that the term 'visual saliency' was how it is described amongst neuroscientists, you wouldn't find all the good papers on the subject.
- **Introduction**
  Next is the introduction. This describes the state of the art of what is known on the subject, what previous research has found and why the authors are investigating the research question. This can be a good section to read if you want an overview of the research on a topic, although, of course, newer papers will provide a more up-to-date overview than older ones.
- **Methodology or procedure**
  Then are sections on how the research was done. This usually covers the type of people who were recruited to take part in the research, the materials that were used in the research, and what was actually done. Often this can be the most technical part of a paper (particularly if it is

discussing the statistical methods used to understand the data). There are two main reasons why these sections are included. First, so that other people can do the research for themselves, if they want to check whether they get the same results (this is a nice idea in theory, although almost never done). Second, so that you have, in theory, all the information you need to be sure that the researchers conducted a fair test, and that the conclusions they are drawing from the results are reasonable.

- **Conclusion**
  Finally there are sections on the authors' conclusions. Next to the abstract and introduction this is often the most useful section of the paper, and if you are trying to get a quick overview of a paper I would recommend that you read the abstract and conclusion sections first, then the introduction.

A particularly useful and information-rich style of paper is the review paper. Review papers are summaries of many experiments on one subject. They summarize what has been found on a particular question. They are useful because often findings from one research paper are not always replicated by other researchers. There can be particular details of the way a test was set up that were responsible for the results obtained. Set up the test in a different way, and you will get different results. This is called the replicability or repeatability of the results.

One of the useful features of papers is that they reference other relevant papers. In other words they can act as a way to discover more research.

If there is a particular paper referenced in this book that is of interest to you, or that is relevant to a design concern or problem you're trying to solve, there is a way to discover if there is any new similar research. Go to Google Scholar (https://scholar.google.co.uk/) and copy in the exact title of the paper in quotation marks. The paper should appear in the search results. Underneath there should be a link that says 'Cited by' and then a number. This is the number of research papers that have referenced that paper. Click on that link for a list of all those papers. Then, to find the most recent ones, there will be a time range for searching, on the left side of the screen. At the top it says 'Custom range' and underneath it will allow you to search for papers that were published after a specific recent year. At the bottom of the left-side menu there is also an option to 'create alert' – this will enable you to receive an automatic e-mail whenever a new paper is published that mentions the paper you are interested in.

Limitations of published research

Just because a paper has been published does not mean it is flawless. There may be logical errors, the experimenters may have failed to control for other factors that could have been affecting their results, or they may simply overstate their claims. The process of reviewing papers for submission is designed to help filter out these problems, but it is not perfect. Also, most research does not get tested or replicated. Research can be expensive and time consuming and we can be more certain about results that have been replicated by another team of researchers. Nevertheless, published papers are an invaluable resource if you read them carefully and with these potential weaknesses in mind.

It is also worth mentioning that most commercial neuro design research is never published in academic journals. A lot of research on design is now being run by commercial companies for their own practical use. They don't usually have the time or inclination to publish their findings in journals.

Proscriptive versus descriptive findings

There is a difference between findings that are proscriptive and descriptive. Sometimes research shows a relationship between two variables – ie particular types of design produce a particular effect in viewers – but without being able to explain why. All they have done is describe a particular effect. This is a descriptive finding.

It is better when actual research results confirm a theoretical prediction: in other words the researchers are able to show *what* happened and explain *why* they think it happened. This type of research gives us more confidence that we can copy the same design technique and get predictable results. This is a proscriptive finding.

The strongest type of research finding is one that has both this type of theoretical support and has also been replicated: more than one study has obtained comparable results. By combining these two factors we can get a good rule of thumb for rating any particular neuro design idea (see Figure 11.2).

Figure 11.2 Combining theoretical and experimental support

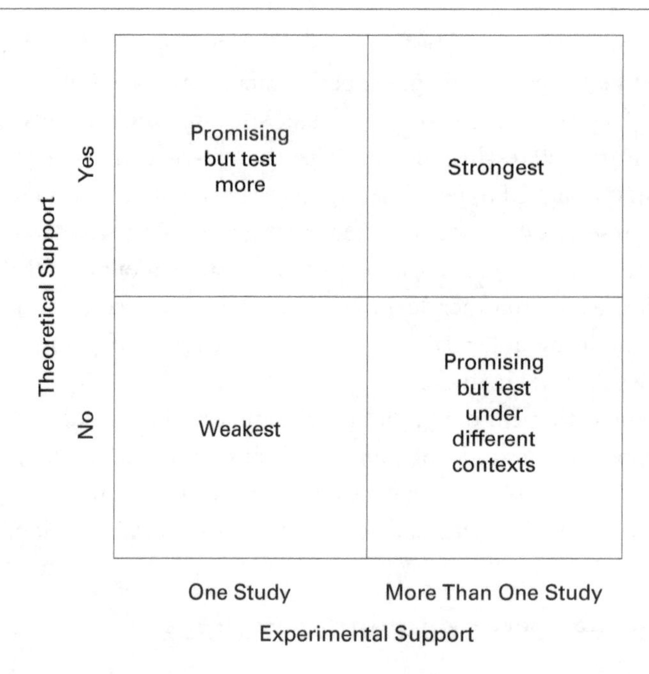

Most neuro design results are trends rather than absolutes

Human beings are not mechanical machines with buttons you can press that create predictable responses. Most research on the psychology and neuroscience of how we respond to visuals shows tendencies rather than absolutes. In other words, a study might find that most people respond in a particular way, or that more people respond in a particular way to design A than to design B. Not everyone will respond in exactly the same way. With most design techniques the best we can hope for is that most people will respond positively.

The benefit of using neuro design techniques is to improve a design, or increase the chances that you will get the effect you want, but it will not guarantee that every single viewer will love it.

Conclusion

Testing images can be very useful but it also should be kept in perspective. Google A/B-tested 40 different shades of blue on their advertising text links in order to discover the optimal shade. The testing reportedly resulted in an extra US$200 million in ad revenue.[3] Google's reach – its ability to A/B test amongst millions of users – can seem too extreme for some designers, however. For example, the former lead of visual design at Google, Douglas Bowman, left his position, tiring of this approach to testing tiny details of design. He later wrote: 'I had a recent debate over whether a border should be three, four or five pixels wide, and was asked to prove my case. I can't operate in an environment like that. I've grown tired of debating such minuscule design decisions. There are more exciting design problems in this world to tackle.'[4]

'Testing can only tell you so much', writes technology journalist Cliff Kuang, 'and it often only reveals that people only like things that are similar to what they've had before. But brilliant design solutions convert people over time, because they're both subtle and ground breaking.'[5]

Testing can help with incremental evolution, clearing an already defined pathway to make it easier to walk down. However, it doesn't tell you whether it is the right pathway to be taking in the first place. Participants come to tests with preconceived ideas about designs they are familiar with. Testing designs that are close to what they are familiar with gives them a built-in head start over more radical ideas. There is not the time within a test session to acclimatize people to a more radical idea, so there is always the risk that testing supports existing ideas more than new approaches. Whilst it may seem paternalistic, sometimes designers may effectively need to say 'I've done the hard thinking for you, and I've discovered a design that you might find strange to begin with, but trust me you'll find it far better in the long run.'

Summary

- Whilst asking people what they think of a design can be useful, you need to be aware of the limitations of this technique and how it can mislead you.
- A/B testing is a powerful and comparatively inexpensive research technique. It works well on websites, and for testing actions (eg whether people click on a link, sign up for more information or buy a product).
- There is now a range of neuroscience research techniques available for medium- to high-budget research projects.
- Eye tracking is good for showing where people look, which design elements they notice (and those they don't), and in what order and for

how long. However, it does not measure what people think or feel about a design.

- Facial action coding is good for measuring the six universally recognized facial emotions, particularly with stimuli that change over time, such as videos or the experience of surfing a website.
- Implicit response testing is good for measuring the associations that a design evokes. It works better with still images than videos.
- EEG/SST and fMRI testing are more expensive measures of things like attention levels, made directly from the brain.

Notes

1 Adams, S (2013) *How to Fail at Almost Everything and Still Win Big: Kind of the story of my life*, Penguin, London.

2 http://www.science20.com/news_releases/blind_people_use_same_emotional_expressions_because_they_are_innate_not_learned_study (last accessed 25 August 2016).

3 https://www.theguardian.com/technology/2014/feb/05/why-google-engineers-designers (last accessed 25 August 2016).

4 http://stopdesign.com/archive/2009/03/20/goodbye-google.html (last accessed 25 August 2016).

5 http://www.fastcodesign.com/1662273/google-equates-design-with-endless-testing-theyre-wrong (last accessed 25 August 2016).

Conclusion 12

Figure 12.1 In the future, neuro design ideas may be applied to areas such as education, cinema, architecture, video games and fashion

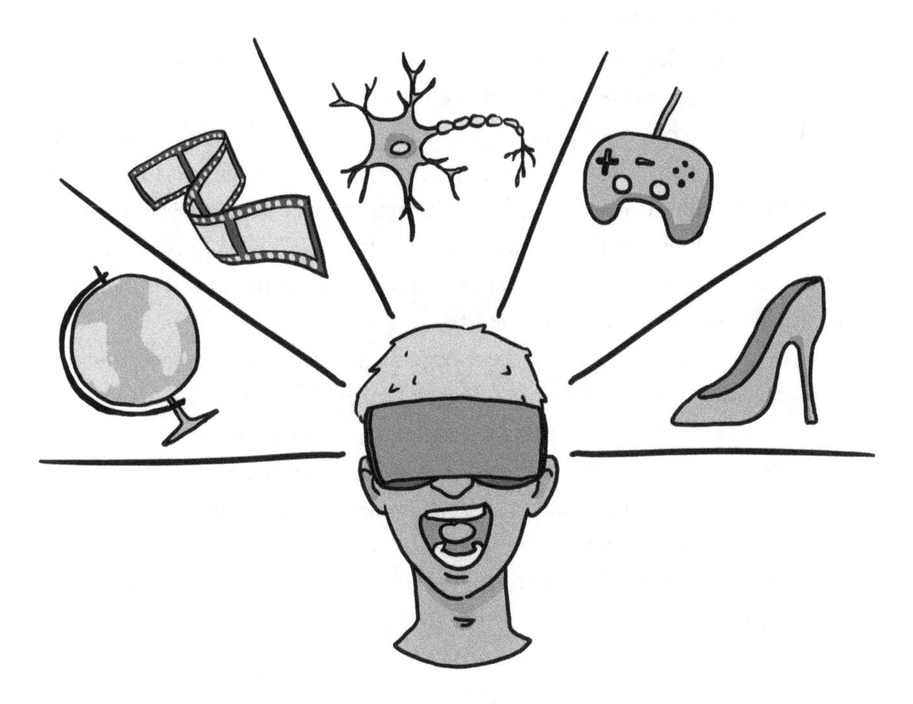

The Glass Bead Game, a novel by Hermann Hesse, set in the 25th century, intriguingly describes how a futuristic society might blend art with science.[1] The novel concerns a game, more intricate than chess, which enables its players to blend together knowledge from the arts and the sciences. The biologist EO Wilson has described something similar, calling it 'consilience', defined as unifying the humanities with the sciences.[2] These two areas are typically divided as separate spheres of study and activity. Yet artists and designers often make use of other areas of knowledge, such as maths, engineering or computer science. Neuro design offers the prospect of – at least partially – uniting art and design with psychology and neuroscience.

Adding neuro design thinking can increase the emotional engagement, attention-grabbing power or memorability of products and services. As we have covered in this book, there is already a range of insights from the neuroscience labs into how to create more effective imagery, and this body of knowledge is likely to increase in the future. In this chapter I describe why I think neuro design will become increasingly used in the years ahead, including how some new technologies could spread its use and the different application areas it might be applied to.

First, by using neuro design I believe organizations can create more effective communications. They will create imagery that better connects with people, but it could also help them to value the role of design more. Adopting a neuro design approach can help elevate the importance of design within organizations. Many organizations in which decisions are ultimately driven by the financial departments have historically adopted rational models for how consumers behave, similar to the AIDA model described in Chapter 1. They may accept that design has a role to play in their success, but because, until now, it has been hard to quantify that role, design ends up with less authority and respect than business activities that can be tracked on a spreadsheet. By demonstrating a quantifiable influence of design it can be better recognized by decision makers.

Companies need to drop the outdated assumption that consumers are totally rational. Those companies that better understand effects such as priming, visual saliency, first impressions or behaviour economics nudges are likely to outcompete those still operating on a completely System 2, rational model. Their communications will be closer to matching how consumers really take in and respond to information. Until now, when trying to test images we have had to rely on what people say. Now we have a whole suite of tools that can capture more intuitive, gut-feeling responses.

However, whilst neuro design holds the prospect of humanizing the more rational businesses, is there a danger of it dehumanizing and overrationalizing the process of design itself? Great design can touch us emotionally, and speaks to our irrational, non-conscious minds. It is not an entirely systematic process. Sometimes the right solution to a design problem might be one that breaks some of the rules. Sometimes following a rule does not bring the result you expect. Equally people's preferences change with the mood of the times, fashion and current style. Could reducing design to a series of rules, or even programming software to create designs, take the humanity and heart out of the process?

This may be a concern if a neuro design approach is followed too rigidly. However, I believe there are a couple of reasons why this need not be a problem.

First, as mentioned at the beginning of the book, neuro design is probably best used as a tool to enhance designers' own intuition and skills. Design education already imparts various rules and techniques to designers. Neuro design just enriches these. Unlike art, design usually has a goal. It needs to create something usable, or create a specific effect in those who view it. Neuro design just helps the designer by giving them more possible ideas for how to reach that goal or how to solve specific design challenges.

A human designer's own creativity and intuition remain the source of good design. Neuro design is at its best when it is helping to tweak and adapt designs, not telling you what you should design in the first place. Neuro design can point to some types of content that might be more successful for particular goals – for example, the knowledge we have about the types of image subject that are more likely to go viral. However, in general, neuro design recommendations are more about how to style an image than what image to create in the first place. In one sense, neuro design is dependent on designers; without them there would not be any images to test! In this way, designers themselves become part of the research process, experimenting with new styles to see how people react to them.

Second, it is probably inevitable that some areas of design will become automated. With the ever-growing need for new designs in displays and webpages, and the need to reconfigure them quickly for different requirements (such as personalizing the look of a webpage for the visitor), it is already impractical for designers to create every design anew each time. There are not the time and money resources for this. Instead think of this category of design as neuro-inspired templates. Design templates are already used extensively in web design, for example. Computer software will likely take pre-existing design elements created by humans, be they drawn images or photos, and mix and mash them in new ways to achieve particular effects. For example, San Francisco web company The Grid (https://thegrid.io/) offers artificially intelligent software for creating websites. You tell the system the basic style of site you want (eg professional or casual) and your priorities for the site (eg more sign-ups, media plays or sales) and it will automatically design and build the site for you. Also there are smartphone apps that will take a photo and render it in different graphical styles, such as making it look like a painting. These are just two examples of how certain elements of design are becoming automated.

How industry is warming to the neuro design approach

My long-time colleague in developing and advocating neuro design research, Thom Noble, is a veteran of both the marketing and market research industries. He has seen how the neuro design approach has been adopted by creatives more enthusiastically than traditional market research:

> *Creative teams have long had difficult relationships with traditional marketing research – seen as an unwelcome hurdle that undermines, emasculates and compromises their work. And one that in any case has little credibility and validity in their eyes. How could it be otherwise? After all, traditional research measures logical/rationalized responses – not those non-conscious thoughts and feelings that the creatives are striving to evoke.*
>
> *The insights from science-led approaches are felt to engender a greater understanding of creative* cause-and-effect *triggers and thus help develop what I would call* 'augmented intuition' *amongst creative teams: a more acute sense of how to evoke the intended respond patterns with the target audience. Far from* constraining *creativity, my view is that used in this way, science is* liberating *it!*

Yet separately from concerns about neuro design possibly overrationalizing and dehumanizing design are concerns over possible ethical misuses of these techniques. As we have learned more about human psychology, and as competition for consumers' attention has intensified, many designers have discovered ways to become more persuasive and create more addictive experiences. For example, developers of smartphone games use knowledge of the psychology of play to maximize the addictiveness. Everything from the colourful look of the graphics, the physics and motion within the game, the scoring, and the difficulty of each level is tested and tweaked to make the experience as pleasurable, addictive and compulsive as possible. On the one hand this kind of intensifying of the pleasure of design gives users more enjoyment, but taken too far could it be unethical in certain contexts? For example, there are already growing concerns in developed economies about advertising and packaging of highly sugared products that are aimed at children. Equally, as mentioned previously, some countries have forced tobacco companies to remove designs from cigarette packets, aware of the role that design can play in making these products more seductive.

Yet I think these concerns are not unique to neuro design, they are already existing concerns with how we choose to approach advertising and commercial activity in general. Designers have been figuring ways to persuade for centuries; neuro design may supercharge their effectiveness, but it is just an extension of something that was already there.

Applications for neuro design

Design extends into all spheres of life. As well as areas such as web design, advertising and packaging, there are other areas of design that might be influenced by neuro insights and research, as set out below.

Neuro educational design

Neuro design research can give insights into how to make information more memorable and engaging. In the same way that infographics are an effective way to convey complex information in an accessible, graphical way, there is a lot of scope for using design and imagery as educational and learning tools. Scientific and mechanical concepts, for example, are prime topics for being taught graphically. Things like the inner workings of biological cells, or the mechanical workings of engines, benefit from seeing cutaway illustrations or animations. These types of illustrations already exist, but by using neuro design insights many more areas of education could be enhanced by highly engaging, enlightening and memorable imagery.

Neuro cinema

Theatrical films are already tested with audiences, whose feedback is used as a likely indication of a film's success with audiences, often used to re-edit or even reshoot sequences, and to choose which sequences to include in the trailers. Neuro design research could help make this process more effective. Using research tools that track viewers' responses to the movie, moment by moment, researchers could learn which types of visuals are most likely to trigger different emotions. In particular, which images are most likely to lead people to want to see the movie, hence are best used in the trailer. Even in the early stages of planning movies, neuro image research could be used to test which actors are most likely to have the right image to connect with audiences, given the emotions and subject of the movie.

Computer algorithms have already revolutionized movie special effects. Movie artists now use computers as just another tool, helping them to create realistic characters and environments. Neuro design could help with the challenge of making artificially created film imagery appear more realistic. For example, when filming the movie trilogy of *The Hobbit*, the director Peter Jackson had planned to film the gold-coin-filled mountain lair of the dragon Smaug using computer-generated heaps of coins. A computer simulation was built to visualize millions of gold coins piled up, and the correct physics was programmed in (the correct settings for how size, weight and shape of the coins should affect their movement). However, when Peter Jackson cast his expert director's eye on the computer-generated film of the piles of coins moving, it didn't look right to him, and the team of artists couldn't figure out why the movement of the coins looked wrong. Eventually they abandoned the computer-generated imagery and were forced to make thousands of real coins to shoot. When computer-generated imagery should look real in theory, but doesn't in practice, something is missing. That something might be some aspect of physics that has not been properly captured. However, the solution might be a better understanding of the visual cues that tell us whether something looks real or artificial.

Neuro architecture

The way our man-made environment is designed can affect our thoughts and feelings. For example, the 'cathedral' effect is the fact that people feel more creative when they are in rooms with high ceilings. Neuro design can help provide insights that will make buildings and interiors more beautiful. It could also help design buildings that put people in the right state of mind. For example, can workplaces be designed (using techniques similar to the cathedral effect) to help make workers more productive, more creative, and enjoy their time at work more? Can hospitals be designed to be more relaxing and friendly places for their patients? Can children's wards and medical equipment used on children be made to look friendlier and less scary? One hospital covered their fMRI scanner (sometimes an intimidating-looking machine) with colourful patterns to make it feel less scary for patients, and more interesting to lie inside.

Neuro fashion

As fashion is something that often taps into the mood of the times, and social trends, its variability might make it less predisposed for neuro design than some other areas. However, novelty and beauty can be very important for fashion design, and neuro design can help provide insights here. Equally, companies and organizations designing uniforms for their workers can use neuro design research to help create clothes that best convey their company ethos – for example, which colours and styles of clothes are best at conveying expertise, friendliness, authority or trustworthiness. Individuals might use neuro design insights to pick the best clothes to wear for a job interview, or to wear in their photos on online dating sites to appear most attractive!

Neuro video games

The video-game industry is now larger than the movie industry. As computers get faster, and graphics become more realistic, games have the potential to become ever more immersive. Games need to emotionally engage us in order to create fun experiences. Similar to neuro cinema, researching video games can help identify the triggers in games that create the strongest emotional responses. Most complex games try to evoke feelings of thrill or fear, but there is also potential for creating other feelings such as empathy, surprise, awe or amusement. Making games graphics feel more realistic, or picking the best graphics to use in ads for games, are also areas that could benefit from neuro design research and insights.

Neuro design apps for consumers

Smartphone apps for filtering and editing photos are now used by millions of enthusiastic amateur photographers. The rise of social media, messaging apps and personal webpages has driven an interest not only in taking photos but in taking photos that impress, or that others want to share. It is possible that in the near future there will be consumer apps that use neuro design principles to enhance, edit or filter images to increase their likability or shareability.

When a professional photographer takes a photo they are bringing a lot of intuitive knowledge to how they frame and compose the shots, how they light them, the settings they use on the camera, and then, ultimately, which of the many shots they select to be the one that gets shown. At least some of this knowledge can be captured in software and made more accessible to amateur photographers.

The new generation of realistic retail simulations

3D visual simulation has existed for a number of years, pioneered by the industries that had the money and need for it – such as flight simulators for the aviation industry – and more recently we have all seen the advances in realistic simulations in movies and video games.

UK-based Paravizion specializes in helping bring this type of simulation to packaging designers. Their software not only makes 3D simulation quick and easy for designers but then, crucially, allows them to place it in a realistic shop setting (for example, a supermarket shelf display) to see how it looks in context and against the competition. Their team can go into a shop and visually scan it in three dimensions, creating a model that pack designs can then be inserted into.

If a designer wants to see the effect of changing colours, textures or lighting conditions, their system can quickly cycle through different options in real time, showing what they will look like. They can see all of this in the simulated environment of the shop, testing the relative effects of changing these variables.

New screens and formats

A significant driver of neuro design in the coming years will be the proliferation of electronic screens. As it becomes cheaper to manufacture screens, they are proliferating in many new shapes and sizes – from tiny screen displays on electrical products and small smartwatch screens at the low end, to giant immersive IMAX screens at the upper end. In-between there are many new screen sizes and formats that will be available soon.

Just as smartphones have obliged users to learn and become comfortable with new design interfaces and aesthetics, new types of screen media will undoubtedly evolve their own design languages. Just as we view and interact differently with TVs, computer screens and mobile devices, the new devices will have their own quirks and demands that require design insights in order to make them accessible to use, and optimize their formats to be most engaging for users. For example, the small screens on smartwatches are requiring designers to rethink how they display information. How can designs be optimized for such a small screen? What kinds of designs are likely to work best at this size? Each of these new media will need new types of designs both from a viewing and an interaction perspective.

Cheap paper-like screens

Another way that screens are set to proliferate more in our lives is a technology called e-ink paper. Until now we've had screens and we have had paper, two separate media. But e-ink paper is more like a hybrid of the two. It raises the possibility of books that combine the best of printed books with the best of e-books. As we saw in Chapter 8, reading from screens can be an inferior process to reading from paper, as screens lack the physical and intuitive feedback we get from books. However, e-ink paper could solve these problems, whilst retaining the dynamic benefits of screens (ie the ability to change images, and use animation).

Like the magical books in the Harry Potter stories, these books will have the physical look and feel of paper books but will have moving imagery on their pages. As technology writer Kevin Kelly describes it: 'E-ink paper can be manufactured in inexpensive sheets as thin and subtle and cheap as paper. A hundred or so sheets can be bound into a sheaf, given a spine, and wrapped between two handsome covers. Now the e-book looks very much like a paper book of old, thick with pages, but it can change its content.'[3]

Designs that are currently printed on to paper or card, and hence are fixed and static, could become more dynamic. Imagine, for example, packaging in supermarkets being made up of thin screen surfaces. The designs on packs could then change throughout the day, depending on the demographics of the shoppers in-store. Or perhaps depending on the weather outside (eg if it is a sunny weekend the uses of a food or drink product for barbecues or picnics could be displayed).

Virtual and augmented reality

Virtual reality is when screens are worn on a headset close to the eyes to immerse users in a 3D world completely apart from the real environment that they are physically in. Augmented reality headsets are see-through, so that their graphical displays get overlaid on top of the user's view of the real room that they are in. They are a bit like the 'heads-up' displays that fighter pilots have had for years. Both technologies are currently poised to make the transition from the development lab into the world of consumer products, and therefore could become popular new screen formats in the coming years.

Virtual reality requires making the simulated world on the screens feel real, even though the body is in a completely different environment. In many users this can result in a feeling of motion sickness, as the physical feedback from their head and body does not match the visual feedback from their eyes. One visual trick used to overcome this is by placing an artificial nose in the centre of the screen, comparable to the view of your own nose that is constantly within your visual field (albeit that we consciously tune out). By inserting this, as the person moves their head around, the subtle view of the nose in the bottom of their visual field also moves around, helping to give the experience a more physically natural feel. Another problem with virtual reality is that a person might want to walk longer distances within the simulation than the physical room they are in allows for. The user does not want to be surprised by suddenly hitting their head on the wall of the real room when it looked like open space in the simulation! The solution to this problem involves some clever trickery to fool our visual system. Each time the person turns their body in the room they are walking in, the simulated world turns a little bit less. It is not immediately perceivable, but it tricks the person into thinking they are walking longer distances in the simulated world whilst in the real room they are moving around in circles, avoiding the walls. Many such visual tricks will undoubtedly be developed as designers work with creating virtual reality simulations.

Augmented reality (AR) offers the prospect of integrating computer-generated displays and environments into our everyday world around us. We already have many physical displays in our real world – from computer and mobile screens to posters and billboards – but augmented reality will multiply these dramatically. By putting on a pair of AR spectacles in the near future, information and graphics will be hovering and overlaid onto the world around you constantly. As you walk down the street you will see real-time information on things like incoming text and e-mail messages; weather forecasts and calendar entries pop-up in front of your view, as appropriate to the time of day and according to how you have set the display. As you walk past shops, personalized offers and recommendations will make themselves known to you via graphical ads that appear to be overlaid onto the walls or windows of the shop.

Finding the best ways to display information graphically in AR will require a lot of research and understanding of visual psychology and neuroscience. What is the best way to integrate information into people's daily life without overtaxing them, or making them bump into objects around them? What are the best ways of portraying complex information in AR displays?

Taken together, these new technologies mean that screens are going to blossom everywhere in our lives, increasing the exposure of designed images of all kinds. As screens delivering information proliferate everywhere, how we decide how to allocate our attention becomes a more pressing challenge. The number of webpages, articles, videos and images to look at is growing exponentially, yet our attention spans and time are finite resources. Whether at work or in our leisure time we are all likely to be dealing with increased flows of information.

Here too the neuro design approach can help. We know that images can help make complex information more easily understandable and quicker to consume. We also know that when images are created that do this (make information easier to understand than expected) they tend to be more successful and more liked. Therefore it should be an obvious target for designers to develop better ways to simplify the complex with images. The skills behind creating good infographics are likely to become in increasing demand.

Take advertising, for example. Currently advertising spending is a bit like a shot in the dark, unsure of whether it will reach its target. The famous lament of the advertising industry to date is that they know that half their ad spending is wasted, they just don't know which half. With greater ability to track what ads and images each person has seen, with their actual spending behaviour – both online and in-store – a greater understanding can be built of how exposure to images can lead to purchasing. For example, at the moment little is known about how many times it takes a consumer to see an image of a brand or product before they are likely to recall or recognize it – or how factors like the context, environment or time of day affect this. After a certain number of exposures of an image – such as an ad or a pack design – it may become so familiar to the viewer that it starts to lose some of its appeal and becomes boring. This point can be tracked and measured and, when reached, the image can be simplified or changed to appear more intriguing again.

More data for finding patterns

An increased ability to track people's exposure to images and responses to them online will generate a lot of new data for neuro design researchers.

Making sense of this, and searching for meaningful patterns (ie what types of images create what types of responses) will be a challenge. Large data sets alone do not necessarily automatically yield insights. If a data set is large enough, correlations between causes and effects can be found that are purely random. In other words, if you look for patterns long enough, you will find some that are random. As Rory Sutherland, vice-chairman of the Ogilvy group in the UK, says: 'the more data you have, the more gold is contained therein... but at the price of an even greater volume of false gold: spurious correlations, confounding variables and so forth. Constructing an inaccurate but plausible narrative is much easier when you can cherry-pick from 50 pieces of information than from five.'[4]

This is why proper controlled tests, combined with a prediction of what you expect to find, are still valuable. Otherwise you could fool yourself into thinking you have found patterns that are actually just mirages.

However, theoretical knowledge can be combined with large data sets. If patterns emerge from online data that make theoretical sense, and can then be checked with future measurements, this can be a valuable research approach.

Computer vision

Using software to understand visual inputs from cameras is a rapidly growing field of research. Computer or machine vision has a number of applications. For example, self-driving cars require very sophisticated visual recognition software. There are also many applications for recognizing faces from photos, videos and cameras, not least of which are the various security applications for finding a face in a crowd. Affective computing is a related field, seeking to recognize human emotions from inputs such as camera images of people's faces. Affective computing will help our computers and everyday gadgets better understand us, by recognizing and tracking our feelings. Similar to facial action coding, this area of research could also be applied to neuro design research, potentially extending the current facial action coding measurements into new emotions, beyond the six universally recognized facial emotions.

Equally, as the number of images online explodes, software will need to get better at recognizing, tracking, searching, indexing and organizing images. Already the free Google Photos service can analyse the photos its users have uploaded and can recognize and tag objects – such as landscapes or buildings – and even recognize what is happening in a photo, such as if it is a birthday party or a music concert. The software can even use backgrounds or landmarks to recognize where a photo was taken (if geotagging information is not available).

The web contains many markers for how much people like images. Photos get likes on Facebook, get repinned on Pinterest, receive hearts on Twitter or likes on Instagram. Information on the likeability and virality of images is a growing free resource available to researchers. When combined with machine-vision tools for analysing images, the potential for developing ever-increasing sophisticated understanding of what drives people's reactions to images is clear.

As software gets better at decoding camera inputs this will have spin-off applications for neuro design. The software for decoding images and photos could become a powerful research tool for neuro design. We can analyse vast databases of images with software in a way that was never possible before. Prior to this kind of software, it would have taken many thousands of hours for human researchers to trawl through images and code them for their various visual features. Now it can be done almost automatically.

What neuro design still needs to learn

Neuro design is currently still a young field. There is still much to learn. In particular there are four areas that need more research, as set out below.

Effects over time

Most testing is done as a one-off exercise. People are shown designs or images and react to them. Less is known about how their reactions change over time. For example, what are the effects of seeing an image multiple times over the space of weeks or months? Are there patterns in how people's responses evolve over time? How does people's ability to recall or recognize various types of images fade over time, and how is this affected by the number of exposures they have had to the image? All these questions will require further research.

Understanding individual tastes

Most of the ideas in this book are aimed at effects that should work across a broad spectrum of the population. These effects are operating at a non-conscious or only semi-conscious level. When it comes to how our brains process the lower-level features of images – their contrast levels, basic features of composition like symmetry, or recognizing emotional expressions on faces – the mechanisms of our minds are more similar than different. However, we know that tastes can vary enormously. Personal experience, the culture we have grown up in, the history of imagery that we have been exposed to in the past, our personalities and so forth all layer on top of those more primitive reaction layers of individual taste. For example, certain cultural references will be recognized by people who have grown up in one culture, but not another, by one age group, but not another.

Understanding more about individual taste in images could help create more personalized experiences. For example, a website that someone visits frequently could map enough information about them to learn their tastes, and then serve up versions of images that are more likely to appeal to them. Web users are already familiar with the content of ads displayed on sites such as Facebook being tailored to their interests (eg products they recently searched for). This would be personalized tailoring of the style of the ads or imagery.

There may also be generalizable patterns in taste that are driven by demographics. For example, we might understand more about how nationality, gender, age and education levels influence our visual tastes. Different fonts, colours, contrast levels and so on could be used in different versions of the same images that are displayed on a website, according to where in the world the user is accessing it from.

Interaction effects

A lot of neuro design findings come from tests in which only one design element is varied at a time. However, in reality there are likely many variables that combine, blend and interact in order to form the person's response to an image. When different effects interact, sometimes one effect might outweigh the importance of another. Sometimes apparently contradictory results could occur. We just don't know enough about this yet. Like a newly developed drug, a particular neuro design effect may work, yet have unexpected side effects.

Ecological validity

Ecological validity is a psychology research term that simply means how accurate a theory is at predicting real-world behaviour. So far, many neuro design theories have been born out of lab research. The way people behave in the real world may sometimes be different to how they respond in a scientific test. Even if the research was not conducted in a lab environment, we know that people's responses can be context dependent.

Understanding responses in real-world environments can be made possible with new research tools: for example, wearables. As people go about their daily lives, wearables, such as smartwatches with sensors, can be used to monitor (with their permission) their emotional responses to what they are seeing. If this data is linked with what they have seen (for example, via time stamps on their web-viewing history) it would be possible to build up an understanding of people's responses to images on a large scale in the real world. Of course, whilst this is possible in theory, in reality it might prove difficult to get enough people to sign up to it (for privacy reasons, for example).

Psychology and neuroscience have already discovered many ways in which different design features can affect our responses. The future holds the prospect of many more discoveries. Commercial design agencies are already touting their knowledge of neuroscience and psychological insights, as a way of differentiating themselves and their ability to find new design techniques and insights. Designers have always needed an understanding of human psychology. Whereas in the past this has largely been intuitive, neuro design will help formalize and enrich it. There may even come a time when designers may fairly be described as psychologists who can draw.

Notes

1 Hesse, H (2000) *The Glass Bead Game*, Random House, London.

2 Wilson, EO (1999) *Consilience: The unity of knowledge*, Abacus, London.

3 Kelly, K (2016) *The Inevitable*, Viking, New York, p 92.

4 http://www.spectator.co.uk/2016/08/how-more-data-can-make-you-more-wrong/ (last accessed 25 August 2016).

APPENDIX: 61 NEURO DESIGN PRINCIPLES

The following is a summary list of the principles, 'best practices' and neuro design effects I've covered in this book. They are grouped according to the chapters they are described in.

Chapter 1: A general principle

1. **The buzz of novelty:** novel choices – even of uncertain value – can activate part of our brain's reward system (the ventral striatum).

Chapter 2: Neuroaesthetic principles

2. **Fractal detection:** our brains can detect fractal patterns in images, and tend to like them as they remind us of patterns in the natural world.
3. **The supernormal stimuli effect:** deliberately exaggerating the visual features – such as shape or colour – that help make something distinctive and recognizable can make the image more appealing and memorable.
4. **Isolation:** isolating an image element makes it easier to see and helps draw attention to it.
5. **Grouping:** our brains classify grouped objects together and assume they share something in common.
6. **The Johansson effect:** we can recognize a living creature simply from the movement of a series of dots placed over it.
7. **Contrast:** things that have good levels of contrast are more easily recognizable.
8. **The peekaboo principle:** partially obscuring something can create the sort of simple visual puzzle that our brains like.
9. **Orderliness:** order in an image (such as multiple lines at the same angle) appeals to our brain's love of grouping things together.
10. **Visual metaphors:** expressing an idea in a design detail – such as a stylized font – helps support the communication of that idea.

11 **Abhorrence of coincidence:** composing an illustration or a photo from a viewpoint such that objects line up in an unnaturally perfect way can feel too contrived.

12 **Symmetry:** we have a natural liking for symmetrical images, especially those that have symmetry around a vertical or horizontal (as opposed to an angular) plane (see also Chapter 3).

13 **Abstract art's isolation effect:** different schools of abstract art manage to stimulate very specific parts of our visual system by focusing on one element – such as movement, colour or form – whilst dampening down variation in all other elements.

Chapter 3: Intuitive design principles

14 **The MAYA principle:** proposed by industrial designer Raymond Loewy, it stands for 'Most Advanced Yet Acceptable', and refers to the idea that in order for a design to appeal to the mass market it must not appear to be so advanced that it appears alien.

15 **Processing fluency:** the ease with which we can understand what we are seeing. In general, the easier it is to understand an image, the more we like it.

16 **The mere exposure effect:** the more often we see an image, the more we tend to like it.

17 **The unexpected familiarity effect:** we find images that seem more familiar or simpler than we expected to be particularly pleasurable to look at.

18 **Propositional density:** aim to convey as many meanings as possible in an image with as little visual detail.

19 **The 'novel, compressible pattern' effect:** our brains like looking at images that are simple on the surface but contain hidden patterns.

20 **The beauty in averageness effect:** our tendency to prefer images of things like faces that are close to the statistical average.

21 **Constructal law:** the theory that many geometric patterns in nature are formed because they help energy move more efficiently.

22 **The golden rectangle effect:** we have a preference for rectangles with a height-to-width ratio of between 1:1.2 and 1:2.2.

23 **The rule of thirds:** people prefer photos composed in alignment with an imaginary grid of lines that divide the image into three vertical squares by three horizontal squares.

24 Image to the left, text to the right: when images and text are combined, we have a preference for images to the left, text to the right.

25 Pseudoneglect: we pay more attention and overestimate visual details in our left visual field.

26 The left-side face effect: we are more emotionally expressive on the left side of our face and hence people prefer looking at that side.

27 Perceptual subitizing: we have the ability to intuitively 'clock' up to three or four visual elements without having to count them.

28 The oblique effect: we find it easier to decode lines that are vertical or horizontal than those at an oblique angle.

Chapter 4: First impressions principles

29 The first impressions effect: we form our basic attitude towards an image within a second of first seeing it.

30 The halo effect: the tendency for our initial judgement of an image (eg whether it is positive or negative) to colour all our subsequent judgements and beliefs about it.

31 The expressivity halo: on first impressions we prefer to see people who are visually expressive – even if they are expressing a negative emotion – over those who are hard to read.

32 The Mona Lisa effect: images that combine details at different levels of spatial frequency can appear different when seen close up as compared to further away or at a glance. This helps explain the ambiguous smile of the Mona Lisa.

Chapter 5: Sensory principles

33 Multisensory integration: the tendency of our brains to blend information together from multiple sources.

34 Synaesthesia: the phenomenon – prevalent to different degrees in people – to automatically link together different sensory qualities, eg to experience letters as having colours, or shapes as having colours.

35 The chromostereopic effect: the tendency for certain colour combinations – such as blue and red – to be hard to look at, due to the way that our eyes process light.

36 The affect heuristic: the tendency for the emotions generated by an image to affect our decisions, even if it seems irrational.

37 Face sensitivity: we have dedicated brain regions for decoding faces and can detect an emotion in a face within one-tenth of a second.

38 Kawaii: the Japanese cuteness aesthetic plays on our brain's natural tendency to find the features of babies – such as large eyes, a large head and chubby limbs – to be naturally cute.

39 Preference for curves: we generally prefer images of curvy objects more than angular ones.

Chapter 6: Saliency principles

40 Visual salience and bottom-up attention: our eyes get automatically drawn to images or areas of images with high visual salience.

41 Mere selection effect: if you merely look at a product you are more likely to choose it.

42 Implied motion: images of arrows can trigger our brain's sense of motion.

Chapter 7: Influence principles

43 Priming: the images we see immediately before or at the time of a decision can influence our behaviour without having to persuade us rationally.

44 Liking is different from wanting: our brains have separate circuits for liking an image versus wanting it. It is generally easier to provoke the desire of wanting than the pleasure of liking.

45 Risk aversion: we have a natural sensitivity to images that trigger the prospect of risk.

46 The mental availability heuristic: our decisions can be biased by information that is easy to visualize. Hence our overestimation of the possibility of dying in a plane or car crash (easily visualized) compared to something like heart disease (harder to visualize).

47 Anchoring and framing effects: our decisions can be influenced by the way that the range of options are depicted.

48 Hyperbolic discounting: our tendency to value rewards now over rewards in the future.

Chapter 8: Screen display principles

49 Reading is harder on screens: even if they are in high definition.

50 **Hard to read = hard to do:** if instructions are in a harder to read font, we assume the task is harder to do.

51 **The doorway effect:** when a scene changes around us – such as when a cut to a new scene is made in a video, or we move through a doorway into another room – we are less likely to remember the information from the previous scene or room.

52 **The Zeigarnik effect:** we are more likely to remember incomplete information than information that has been fully concluded.

53 **The peak-end rule:** the strongest emotional 'peak' and the final moments of an experience disproportionately influence our subsequent attitude towards that experience.

54 **The disinhibition effect:** the anonymity of the web encourages a loosening of social constraints in our behaviour online. For example, people make more unhealthy pizza orders (more toppings and more calories) when ordering online rather than face-to-face.

55 **The touchscreen effect:** when people are able to interact with a product on a touchscreen it feels more like touching it in real life than when merely seeing it on a screen.

56 **The screen-size effect:** people respond more intuitively and irrationally to rich media on large screens than text on smaller screens.

57 **Central fixation bias:** in displays with a central position our eyes are drawn there; when there is an array with no absolute centre, our eyes are drawn to the top left.

58 **Horizontal viewing bias:** we find it easier to scan visually from side to side rather than up and down.

Chapter 9: Viral design principles

59 **Mimetic desire:** seeing someone pick up an object makes us like it more.

60 **Emotional virality:** emotional content is more likely than non-emotional content to go viral online.

61 **Viral images are different from beautiful images:** the factors that make an image viral are different from those that make it aesthetically pleasing.

INDEX

Note: *Italics* indicate a Figure or Table in the text.